# THE THEORY OF ELECTROMAGNETIC
## FLOW-MEASUREMENT

# THE THEORY OF
# ELECTROMAGNETIC
# FLOW-MEASUREMENT

BY

## J. A. SHERCLIFF

*Fellow of Trinity College and
Lecturer in the Department of Engineering,
University of Cambridge*

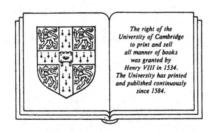

The right of the
University of Cambridge
to print and sell
all manner of books
was granted by
Henry VIII in 1534.
The University has printed
and published continuously
since 1584.

## CAMBRIDGE UNIVERSITY PRESS

*Cambridge*

*New York    New Rochelle*

*Melbourne    Sydney*

CAMBRIDGE UNIVERSITY PRESS
Cambridge, New York, Melbourne, Madrid, Cape Town, Singapore, São Paulo, Delhi

Cambridge University Press
The Edinburgh Building, Cambridge CB2 8RU, UK

Published in the United States of America by Cambridge University Press, New York

www.cambridge.org
Information on this title: www.cambridge.org/9780521335546

First published in the Cambridge Engineering Series 1962
Reissued in the Cambridge Science Classics series 1987
Re-issued in this digitally printed version 2009

A catalogue record for this publication is available from the British Library

Library of Congress Cataloguing in Publication data
Shercliff, J. A. (John Arthur)
The theory of electromagnetic flow-measurement.
(Cambridge science classics)
Bibliography
Includes index.
1. Magnetohydrodynamics.   I. Title.   II. Title: Electromagnetic
flow-measurement.   III. Series.
QC718.5.M36S54   1987   538′.6   86-26840

ISBN 978-0-521-33554-6 paperback

# CONTENTS

# FOREWORD

Arthur Shercliff was born in 1927, and after attending Manchester Grammar School, Trinity College, Cambridge, and Harvard University he joined A. V. Roe Ltd as a graduate apprentice. He returned to Cambridge in 1951 where he started research for a PhD from which stemmed his work in electromagnetic flow-measurement and subsequently this, his first book.

A mark both of his writing and, for those of us fortunate enough to benefit from it, of his research supervision, were the stimulating insights, and the multitude of ideas to be pursued further. This book is no exception and has provided a widely used reference in many parts of the world not only in academia but also in industry. Perhaps the greatest tribute to the author and his work is that an academic monograph of such excellence should also provide the industrial reference on the subject. Shercliff set down with an elegance and clarity the theoretical foundation of electromagnetic flow-measurement. Commercial flowmeters based on this principle had appeared in the early 1950s, but as with most new flowmeter concepts there was an initial conservatism which slowed down their acceptance. Shercliff's book appeared at a crucial stage and provided a stimulus for the work of M. K. Bevir and others who have been involved in the development of the many designs since then. This book continues to be the best overall introduction to the many aspects of electromagnetic flow-measurement. Obviously it does not include the large amount of recent work but the basic ideas are well set out for the newcomer to the subject.

After an historical introduction the second chapter contains the theory for the most common form of the device, where the liquid is of sufficiently low conductivity not to affect the magnetic field. It is this part of the book that has been of particular value to the industry. Ideas may be found there, such as integrating electrodes and segmental electrodes, which are still being developed.

The third chapter deals with liquid metal flows in which the magnetic field and flow profile are affected by the motion of the liquid. Flowmeters operating in this regime have been found mainly in nuclear power industry applications, but it is likely that applications in other processes, particularly continuous casting, will increase.

Chapter 4 covers other devices, such as probes and ship's logs, and other methods of sensing apart from electric potential, such as magnetic flux distortion measurement. Much development has taken place in these areas so that electromagnetic probes and ship's logs are commercially available and flux distortion flowmeters have been developed for nuclear reactor control.

Chapter 5, Shercliff's view of the future of electromagnetic flow-measurement, is still of interest. It opens with the case for the electromagnetic flowmeter which is now well established, and has found a highly respected niche in applications such as slurries, not to mention many more basic ones. Perhaps Shercliff's emphasis on a standardised design needs to be considered again. We have assumed that manufacturers would not wish to be so constrained, but the value to the user of a design which met the requirements set out by Shercliff could be very attractive, and with greatly extended design methods, which have now been developed, a very versatile instrument may be feasible. Both the Appendix with typical magnitudes and also the exhaustive Bibliography are of continuing usefulness.

Of all the valuable ideas brought together in this book, the most formative is, with little doubt, the weight function. This has proved to be an extremely powerful concept indicating the contribution of the flow, at each point in the cross-section, to the output signal. M. K. Bevir developed from this concept the weight vector with the necessary and sufficient condition for the flowmeter to be ideal (i.e. to measure the volumetric flow regardless of flow profile). This condition is that the curl of the weight vector must be zero. Although the theory is now much more developed, the weight function for rectilinear flows is still found to be a useful criterion of performance. This concept has also been applied to the ultrasonic flowmeter and may find wider use still.

Arthur Shercliff was elected a Fellow of the Royal Society in 1980 and it gave him great pleasure that he thereby increased the engineering representation on that body. He died on 6 December 1983 at the end of his first term as Head of the Cambridge University Engineering Department. A tribute to him appeared in the *Journal of Fluid Mechanics*.* For those of us who knew him and for those who have made use of his work and have valued his insight and inspiration, the reissuing of this book will be warmly welcomed and is a fitting tribute to one of engineering's most brilliant researchers.

ROGER C. BAKER
DEPARTMENT OF FLUID ENGINEERING AND INSTRUMENTATION,
CRANFIELD INSTITUTE OF TECHNOLOGY

* *J. Fluid Mech.* (1984) **138**, 431–2.

# PREFACE

Electromagnetic flowmeters have been in use for several decades, while the basic principle dates from the days of Faraday. An extensive literature in journals and reports has grown up, especially since the Second World War, and it has seemed timely to endeavour to present a systematic and unified account of the theoretical basis for the electromagnetic flowmeters that are now in use and also for variations from standard practice. The treatment given in this book is largely theoretical, but the theory is related to practical considerations, particularly in the later chapters. Typical physical magnitudes are exhibited in the Appendix. Exhaustive discussion of the details of design and procedure, such as the choice of the best electronic circuits, would be out of place in this book. However, there is an extensive Bibliography, which makes it easy for information of this sort to be found.

The theory of electromagnetic flowmeters belongs to the subject of magnetohydrodynamics, formed by the combination of the classical disciplines of fluid mechanics and electromagnetism. Consequently a considerable fraction of this book should be of some interest to those studying magnetohydrodynamics for its own sake or for its technological applications to devices other than flowmeters. As magnetohydrodynamic devices go, the flowmeter may appear a little unglamorous, but it is at least the one which has the longest record of useful service in many areas of science and engineering.

Much of the material of the book is drawn from the published literature but there are several portions that are new. To avoid undue length (and tedium for the practical reader who wants results rather than prolonged analysis) qualitative and order-of-magnitude arguments have been given in many situations where a fuller, more rigorous treatment is readily available in the literature. This comment refers mainly to ch. 3. It is hoped that the qualitative arguments can communicate a physical understanding of the phenomena more graphically than would mere mathematical analysis.

It is a pleasure to record my debts to W. Murgatroyd for first introducing me to the flowmeter problem and to magnetohydrodynamics generally, to the United Kingdom Atomic Energy Authority for supporting experimental work on flowmeters in the period 1951–56 and to W. D. Jackson for several recent discussions.

J. A. SHERCLIFF

CAMBRIDGE
*1961*

# NOMENCLATURE

*Note*: Rationalised M.K.S. units are used throughout this book.

$a$      semi-width of channel (in $x$-direction); internal radius of pipe.

$A$      constant.

$A_m$      Fourier coefficient.

$b$      semi-depth of channel (in $y$-direction); external radius of pipe; inner radius of annular space; semi-distance between electrodes.

$\mathbf{B}$      magnetic flux density vector.

$B$      imposed flux density (in $x$-direction).

$c$      semi-length of pole-faces (in $z$-direction).

$C$      constant.

$C_f$      friction factor, $2a(-\partial p/\partial z)/\rho v_m^2$.

$d$      wall conductivity number, $w\kappa/a\sigma$ or $w\kappa/b\sigma$.

$\mathbf{D}$      electric displacement vector.

$\mathbf{E}$      electric field intensity vector.

$g$      magnet gap.

$h$      length of side of square electrode array.

$\mathbf{H}$      magnetic field intensity vector.

$\mathbf{j}$      current density vector.

$j$      current density magnitude.

$J$      wall current per unit width or length; total current in central conductor.

$k$      root of equation $dk \tan k = 1$.

$l$      representative length.

$L$      distance between electrodes, $XY$; entry length.

$m$      integer.

$M$      Hartmann number, $Ba(\sigma/\eta)^{\frac{1}{2}}$.

$n$      normal distance; odd integer; coefficient of $\theta$.

$p$      fluid pressure.

$Q$      volumetric flow rate.

$r$      polar co-ordinate.

$R$      Reynolds number, $av_m/\nu$.

$R_m$      magnetic Reynolds number, $av_m/\lambda$ or $bv_m/\lambda$.

$s$      tangential distance; settling cell dimension.

$S$      sensitivity, $U_{XY}/LBv_m$, etc.

$t$      time; settling cell dimension.

| | |
|---|---|
| $T$ | settling time. |
| $U$ | electric potential. |
| $U_{XY}$, etc. | potential difference between electrodes $X$ and $Y$, etc. |
| $\mathbf{v}$ | fluid velocity vector (in $z$-direction). |
| $v$ | fluid velocity magnitude. |
| $v_c$ | velocity in core. |
| $v_m$ | mean velocity. |
| $V$ | function of $r$ and $z$. |
| $w$ | wall thickness; velocity perturbation. |
| $W$ | weight function. |
| $x, y, z$ | Cartesian co-ordinates. |
| $z$ | $x + iy$. |
| $z'$ | variable of integration. |
| $Z$ | function of $r$. |
| | |
| $\gamma$ | $a/b$ for annulus. |
| $\delta, \delta_1$ | boundary layer thicknesses. |
| $\Delta p$ | pressure drop. |
| $\Delta U$ | potential difference. |
| $\epsilon$ | permittivity of fluid. |
| $\zeta$ | $\xi + i\eta$. |
| $\bar{\zeta}$ | $\xi - i\eta$. |
| $\eta$ | Cartesian co-ordinate; viscosity of fluid. |
| $\theta$ | polar co-ordinate. |
| $\kappa$ | wall conductivity. |
| $\lambda$ | magnetic diffusivity, $1/\mu\sigma$. |
| $\mu$ | permeability of walls and fluid, assumed equal to $4\pi.10^{-7}$, the permeability of free space. |
| $\mu\sigma\nu$ | magnetic Prandtl number, $\nu/\lambda$. |
| $\nu$ | kinematic viscosity of fluid, $\eta/\rho$. |
| $\xi$ | Cartesian co-ordinate. |
| $\rho$ | fluid density. |
| $\sigma$ | fluid conductivity. |
| $\tau$ | contact resistance per unit area. |
| $\phi$ | magnet flux. |
| $\psi$ | perturbation stream function. |
| $\boldsymbol{\omega}$ | vorticity. |
| $\omega$ | angular frequency. |
| $\Omega$ | velocity gradient, $dv_z/dx$. |
| $\mathscr{I}$ | imaginary part of. |
| $\nabla^2$ | $\partial^2/\partial x^2 + \partial^2/\partial y^2 + \partial^2/\partial z^2$. |

| | |
|---|---|
| c | value in the core. |
| crit | critical for laminar instability. |
| f | value in the fluid. |
| m | mean value. |
| max | maximum value. |
| n | normal component. |
| o | value at $x = 0$. |
| r | radial component. |
| s | tangential component. |
| ult | ultimate value. |
| w | value in the wall. |
| x, y, z | Cartesian components. |

| | |
|---|---|
| dash | (e.g. $v'$) turbulent fluctuation; derivative with respect to $r$; excess magnitude during settling. |
| bar | (e.g. $\bar{v}$) mean value; complex conjugate. |

# Chapter 1

# ELECTROMAGNETIC FLOW-MEASURE-
# MENT SINCE FARADAY

Material moving in a magnetic field experiences an electromotive force acting in a direction perpendicular both to the motion and to the magnetic field. This discovery was one of the foundations of electromagnetism. That it should occur even when the material was fluid did not escape the attention of early investigators such as Faraday, who reported* to the Royal Society of London in 1832 how he had tried vainly to measure the voltage induced across the the river Thames by the motion of the water in the vertical component of the earth's magnetic field. The measurement was made between large electrodes, lowered into the river from Waterloo Bridge. Such signals as he detected were spurious ones due to electrochemical and thermoelectric effects, two factors which can still trouble us when we try to apply the principle of electromagnetic induction to measuring a fluid velocity or bulk flow rate. Faraday's experiments failed chiefly because the river bed would short-circuit much of the genuine signal. However, he lived to hear of Wollaston's † measurements of voltages induced tidally in the English channel in 1851.

A surprising number of years elapsed before the induction of an e.m.f. in fluid moving in a magnetic field was put to practical use. Once again the situations were ones where the fluid occurred naturally. Smith & Slepian‡ in 1917 patented a scheme for finding the speed of ships relative to the sea from measurements of the voltage induced between two electrodes on the hull in the presence of a magnetic field emanating from the ship. An alternating field was proposed to eliminate polarisation. Then Young, Gerrard & Jevons§ studied the tidal motions of the sea in the Dart estuary in Devon and reported in 1920 that the signals induced between fixed electrodes inserted in the water correlated very convincingly with the tidal

---

* Faraday, M. (1832). *Phil. Trans.* 175.
† Wollaston, C. (1881). *J. Soc. Tel. Engrs,* **10**, 50.
‡ Smith, C. G. & Slepian, J. (1917). *U.S. Patent* 1 249 530.
§ Young, F. B., Gerrard, H. & Jevons, W. (1920). *Phil. Mag.* **40**, 149.

motions. Their other experiments with electrodes towed behind a moving vessel were not so successful. Since this work many oceanographers have studied sea currents, waves and tides with the aid of induced voltages, measured either from electrodes attached to ships or on the sea-bed or inserted in the ground on shore near the sea. The e.m.f.s in the water cause detectable currents and voltages in the sea-bed, in submarine cables and in the land adjoining the sea.

Somewhat after these first oceanographic applications, various landlubbers realised that electromagnetic flow measurement could be applied to fluids other than sea water, flowing in artificial environments. It seems that Williams's work,* done apparently out of academic interest, was the first of this kind to be published. He

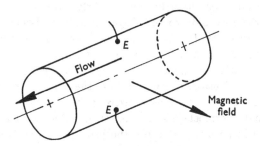

Fig. 1. Transverse-field flowmeter.

performed experiments in which copper sulphate solution was passed along a non-conducting circular pipe, under a uniform transverse magnetic field, as shown in fig. 1. A d.c. voltage was measured between the two electrodes $EE$ and this was found to be proportional to the flow rate. Thus the apparatus constituted a simple type of electromagnetic flowmeter. Williams's experiments also included measurements of electric potential along the line $EE$. He realised that, since the velocity of the fluid would not be uniform over the cross-section of the pipe and would fall as the pipe walls were approached, the induced e.m.f. would not be uniform either. This effect is illustrated in fig. 2 a. The result is that the larger central e.m.f. drives a current back against the weaker peripheral e.m.f. producing a circulation of current much as shown in fig. 2 b. In consequence, the voltage measured across $EE$ is not the sum of the e.m.f.s along the line $EE$ but is less by an Ohmic resistance drop. To put it another

* Williams, E. J. (1930). *Proc. Phys. Soc. Lond.* **42**, 466. See also: Regnart, H. C. (1930). *Proc. Univ. Durham. Phil. Soc.* **8**, 291.

way, the slow-moving fluid partially 'short-circuits' the e.m.f. induced in the fast-moving fluid. This might be expected to be an undesirable complication and to cause the calibration of such a flowmeter to depend on the velocity distribution and the conductivity of the fluid. By a fortunate mathematical fluke, the very reverse is true, provided only that the velocity distribution is symmetrical about the centre-line of the pipe—as it might reasonably be expected to be. It then turns out that the voltage which appears across $EE$ is just the same as it would be if all the fluid were travelling at a uniform speed equal to the previous mean value. In this case the voltage across $EE$ would obviously be independent of the fluid's conductivity because all the induced e.m.f.s would be uniform and no circulating currents or Ohmic losses would occur. Thus in the

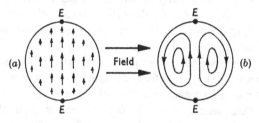

Fig. 2. (a) Electromotive forces; (b) currents induced in flowmeter.

original case, granted symmetry of the velocity profile, the voltage across $EE$ is independent of the detail velocity distribution and the conductivity of the fluid. We are here restricting ourselves to the case of pipes with non-conducting walls. Williams was aware that this result was likely to hold and succeeded in proving it for particular velocity distributions (although his analysis contains mathematical errors). The general proof and extension of this result appears in ch. 2.

Williams also remarked that, if the magnetic field were strong enough and the fluid were a much better conductor of electricity than the electrolyte he was using, the circulating currents would be so intense that significant forces would be exerted on the fluid by the interaction of the currents with the magnetic field.

As it happened, the physiologists, beginning with Fabre's proposal,* were the first to use electromagnetic flowmeters as measuring instruments, soon after Williams's work. Some instrument capable of

* Fabre, P. (1932). *C.R. Acad. Sci., Paris*, **194**, 1097.

recording instantaneous blood flow in arteries was required and the electromagnetic flow-measuring principle provided an attractive solution. The great advantage was that an electromagnetic flowmeter has a response rapid enough to indicate the details of blood-flow pulsations. Moreover, the blood could be kept entirely enclosed and free from contamination. Other advantages were that very little opposition to the flow need be presented by the meter and that in so far as the meter is a *linear* device (i.e. induced voltage is proportional to flow rate), it gives a signal whose mean may be directly interpreted to give the mean of a pulsating flow. These same advantages applied equally or with more force in the technological applications of electromagnetic flowmeters that came later.

Blood proved to be adequately conducting for induced voltages to be measured, particularly if an alternating magnetic field was used in order to eliminate polarisation at the electrodes. Since Fabre many physiological applications of electromagnetic flowmeters have been described in the literature, notably by Kolin. The blood has either been piped to and from a meter, external to the body, or has been metered by a skilfully designed flowmeter which could be inserted round an intact blood vessel in the body.

In 1941 Thürlemann* gave the first general proof of the result, mentioned earlier, that for a given total flow rate but any symmetrical velocity profile the voltage induced across a circular pipe with non-conducting walls is the same as if the fluid velocity were uniform. The simplicity of this result has tended to make people regard the electromagnetic flowmeter as an absolute instrument, entirely insensitive to irregularities of the velocity distribution. This is an erroneous view, as will be seen later.

The next development in the history of the electromagnetic flowmeter came as a result of the advent of the power-producing nuclear reactor after the Second World War. The problem of extracting heat from the more compact types of power reactor, particularly the fast reactor, is so formidable that engineers were led to the use of liquid sodium as a coolant, despite the technological and metallurgical problems involved. Bismuth is another liquid metal that is used in power reactors where the uranium fuel is dissolved in the liquid metal. The practical difficulties in the use of liquid metal circuits have been overcome, with the result that liquid metal technology is established as standard practice, to be ranked with steam technology and other branches of advanced plumbing.

* Thürlemann, B. (1941). *Helv. Phys. Acta*, **14**, 383.

One of the problems of employing liquid metals has been the development of flow-measuring devices capable of meeting the stringent conditions that prevail in or near a power reactor. The device must be able to withstand corrosion by radioactive molten metals at high temperatures. It must need no maintenance and be remote-indicating, in view of the radioactivity, and must have a stable and easily established calibration. All these qualities, and more, are possessed by the electromagnetic flowmeter, which has now become a standard fitting in liquid metal circuits such as those in the Dounreay fast reactor in Scotland. It is normal to use a d.c. field as polarisation gives no trouble with liquid metals.

Nuclear reactor technology also entails the handling of other radioactive fluids, both in the preparation of fuel and in the processing of spent fuel. The measurement of the flow of noxious liquids in these and other situations provides further applications of electromagnetic flowmeters. But even when the fluid is not unpleasant to handle the electromagnetic flowmeter is often competitive with alternative flow-measuring devices, provided only that the fluid conductivity is adequate. With modern electronic techniques* the flow of fluids with the conductivity of distilled water or even less may be measured with an electromagnetic flowmeter. Applications of the device to a wide variety of fluids have been reported in the literature (see Bibliography). The fact that its indication is electric renders it particularly suitable for incorporation in automatic control loops. Various commercial organisations on both sides of the Atlantic now market standard designs of electromagnetic flowmeter complete with the associated electrical equipment. The claims of the sales literature should be read very critically, however.

Another family of electromagnetic devices has been nurtured by the nuclear power reactor. These are the various types of electromagnetic pump for liquid metals. Ever since Ritchie† and his contemporaries it has been known that conducting fluids can be propelled by subjecting them to perpendicular electric currents and magnetic fields. Electromagnetic pumps have become established practice in liquid metal circuits for the same reasons that favour electromagnetic flowmeters.

We have already remarked the fact (which Faraday knew) that the currents induced in a sufficiently conducting fluid by its motion

---

* Lynch, D. R. (Dec. 1959). *Control Engng*, 6, 122.
† Ritchie, W. (1832). *Phil. Trans.* 294.

in a magnetic field could interact with this field to have significant effects on the fluid motion. At the same time the induced currents could have a significant effect on the magnetic field. This two-way interaction between the fluid motion and the electromagnetic field can occur with liquid metals. Its implications for electromagnetic flowmeters form the contents of ch. 3. The study of the motion of electrically conducting fluids in magnetic fields has in recent years grown into a subject in its own right, variously termed 'hydromagnetics', 'magnetohydrodynamics' and other compounds expressing its mixed parentage. It has repercussions in astrophysics, geophysics and many branches of technology, of which the most important is controlled thermonuclear power. In this case the conducting fluid is an ionised gas or plasma. Other potential applications of magnetohydrodynamics to plasma technology are electromagnetic plasma rockets for space propulsion and magnetohydrodynamic generators for the direct conversion of the energy of hot gases into electrical power. There is an obvious analogy between ordinary electric motors and electromagnetic pumps and rockets, just as there is one between ordinary electric generators and electromagnetic flowmeters and magnetohydrodynamic generators. The flowmeter is more strictly analogous to a tachometer of the generator type since negligible power is usually drawn from it.

Though this book is mainly concerned with electromagnetic flowmeters, a considerable part of the theory has some relevance to the other applications just mentioned, particularly the contents of ch. 3.

The electromagnetic induction principle has also been applied to velometry (i.e. the measurement of *local* fluid velocity as distinct from the total volumetric flow rate along a pipe) by having two closely spaced electrodes inserted in the fluid, spanning the region of interest, with a transverse field applied by a magnet external to the fluid or by a small one immersed in the fluid. It might be possible to apply this technique for velocity measurements in ionised gas in the vicinity of hypersonic missiles or in the various plasma devices.

So far only one geometrical configuration for electromagnetic flowmeters has been mentioned, namely, the Williams type in which the fluid passes along a circular pipe under a transverse field with two electrodes at the ends of a diameter of the pipe at right angles to the field and the flow, as shown in fig. 1. This is far from being the only possibility, even among flowmeters which rely on a measurement of induced voltage. In later chapters we discuss electromagnetic flowmeters in which some other physical quantity is measured.

6

It turns out that a pipe of rectangular cross-section has advantages over a circular pipe for a flowmeter in some circumstances, although a circular one is stronger and is more easily made and incorporated in piping circuits.

A more drastic variation from the Williams configuration is to abandon the externally imposed transverse field. It is obviously necessary that whatever field is provided should have a component perpendicular to the fluid velocity if there are to be e.m.f.s induced. If the fluid motion is to be along a pipe of some kind, a possible configuration is that where the magnetic field lines are circles concentric with the axis of the pipe. Such a field results if an exciting current is caused to flow in a direction parallel to the fluid motion,

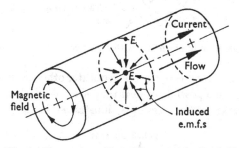

Fig. 3. Flowmeter with axial current in the fluid.

either in the fluid itself or in a solid central conductor, with the fluid flowing in an annular passage. The first reference to this configuration with the current flowing in the fluid was by Kolin,* who pointed out that this meter too can have the property that for a given flow rate the induced voltage is independent of the exact velocity distribution provided that this is symmetrical about the axis of the pipe. Figure 3 illustrates this case. The output voltage is measured between the electrodes $EE$, one at the centre of the pipe, one at the wall. The peculiar advantages and shortcomings of this and other axial-current meters are treated in subsequent chapters alongside those of the more orthodox transverse-field meter.

In selecting configurations suitable for electromagnetic induction flowmeters one must avoid those where the induced e.m.f. is entirely lost or short-circuited. An example of this type would be one in which there is a radial field. Figure 4 shows how in this case the circulating current dissipates the e.m.f. in Ohmic losses. But even

* Kolin, A. (1956). *J. Appl. Phys.* **27**, 965.

in a properly designed flowmeter some short-circuiting is inevitable. Figure 2 has already illustrated how short-circuiting of the induced e.m.f. can take place via the slower-moving fluid near the walls. Any real flowmeter will be finite in length in the flow direction and so there will be short-circuiting at the ends of the meter where the transverse field and hence also the induced e.m.f. fall to zero, as Hartmann* realised in 1937. Conducting walls can also short-circuit some of the output signal of a flowmeter.

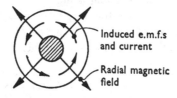

Fig. 4. Cross-section of flowmeter with signal completely short-circuited.

If flow along a pipe is not a pre-requisite, other geometrical configurations become possible even if they are not particularly advantageous in general. One example, illustrated in fig. 5, would be

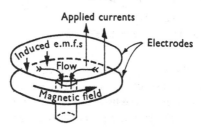

Fig. 5. Radial flowmeter.

provided by taking two parallel circular plates for electrodes and causing the fluid to flow radially between the plates in the presence of a circular magnetic field induced by a uniform current flux between the plates. The voltage between the plates is independent of distance from the centre in this case, but the induced and resistive contributions to it could only be distinguished by stopping the flow and observing the resistive contribution by itself. Outwards flow would decrease the voltage for a given current—the flowmeter tends to self-excite partially—inwards flow the reverse. This is an interesting example of the tendency of a conducting fluid motion to affect an

* Hartmann, J. (1937). *Math.-fys. Medd.* **15**, no. 6.

electromagnetic field, as we have already noted. Alternatively, the field could be excited by a current flowing in an axial straight conductor. The voltage between the plates would then be purely the induced one, but would vary with distance from the centre as the fluid velocity and field strength fell off together.

There may often be situations occurring in existing or projected flow circuits where a given configuration may be exploited for electro-magnetic flow measurement by the addition of suitable electrodes and imposed fields. The more simple and symmetrical the configuration, the better chance there is of devising a meter which gives a reasonably direct and reliable measure of the flow rate. The other extreme confronts the oceanographer who has no control over the field and flow configuration that he studies. As a result many measurements must be taken and analysed before the details of the sea's motion can be established.

# Chapter 2

# INDUCTION FLOWMETER THEORY

## 2.1. Governing equations

The essence of the induction flowmeter problem is this: given the velocity distribution of a conducting fluid and a known or imposed magnetic field distribution, how does one deduce the resulting electrical effects and in particular relate the potential difference appearing between two or more selected points or surfaces (the electrodes) to some characteristic flow quantity such as the mean velocity? The oceanographers have been confronted with a similar problem.

It will be noticed that here we are deliberately ignoring the two other aspects of the motion of a conducting fluid in a magnetic field, namely the effect of the induced currents on the field distribution and the effect on the fluid motion of the electromagnetic body forces associated with these currents. These two effects are significant only with liquid metals and will be discussed in ch. 3.

The fluid is assumed to be non-magnetic, with the same permeability $\mu$ as a vacuum. As a result the fluid can suffer body forces and affect a *steady* field distribution only by virtue of its motion and not by its mere presence.

The fluid is furthermore assumed to obey Ohm's Law, so that current flow is proportional and parallel to the electric field referred to axes moving with the fluid, the conductivity $\sigma$ being isotropic and unaffected by the magnetic field or the fluid motion. The Hall effect is thus excluded. We shall also assume that variations of conductivity and thermoelectric effects due to non-uniformity of temperature are negligible. Currents due to convection of charge by the fluid motion are neglected because such charge as is necessary to account for any divergence of the electric displacement in the fluid is always very small. In electrolytic conductors the effective conductivity between two points can be markedly reduced by transverse fluid motion because of the low mobility of ions, as has been reported by Eskinazi* and others. We shall ignore such effects in our analysis, though their in-

---

* Eskinazi, S. (1958). *Phys. Fluids*, **1**, 161. See also: Hogan, M. A. (1923). *Engineering, Lond.*, **115**, 66. For a good survey of effects peculiar to electrolytes see Wyatt, D. G. (1961). *Phys. Med. Biol.* **5**, 289, 369.

fluence on flowmeter performance could be significant. In electronic conductors like the liquid metals all the above assumptions are wholly realistic.

Thus we take current flow in the fluid to be governed by Ohm's Law in the form

$$\mathbf{j} = \sigma(\mathbf{E} + \mathbf{v} \times \mathbf{B}), \qquad (2.1)$$

in which $\mathbf{j}$ is the current density vector, $\sigma$ the fluid conductivity (a scalar), the reciprocal of resistivity, $(\mathbf{E} + \mathbf{v} \times \mathbf{B})$ is the electric field *relative to the moving fluid* (relativistic terms being negligible), where $\mathbf{E}$ is the electric field in the stationary co-ordinate system, $\mathbf{v}$ is the fluid velocity and $\mathbf{B}$ is the magnetic flux density. The term $\mathbf{v} \times \mathbf{B}$ represents the e.m.f. induced by the fluid motion, while $\mathbf{E}$ is due to charges distributed in and around the fluid and to any variation of the magnetic field in time. A full nomenclature list appears at the end of the book. Rationalised M.K.S. units will be used throughout.

If either or both of the fluid velocity and magnetic field are changing with time, the current induced by the fluid motion will also be variable. Then, if the frequency of change is at all high, the induced current and electric field distributions will be seriously upset by self-inductance. Even if the fluid is stationary the distribution of a time-varying imposed field will be perturbed by self-inductance. This is the familiar 'skin-effect', which in extreme cases excludes the imposed magnetic field from the bulk of the fluid, so rendering electromagnetic flow-measurement impossible. The importance of all these effects which complicate the operation of an induction flowmeter is measured by the same criterion: they can all be ignored if the quantity $(\mu\sigma\omega)^{-\frac{1}{2}}$ be large compared with a representative length scale of the flowmeter such as the pipe diameter (see Appendix for typical magnitudes). Here $\omega$ is the frequency of change of either the fluid velocity or the magnetic field, as appropriate. We shall henceforth assume that $\omega$ is small enough to satisfy this criterion so that the self-inductance effects may be ignored and we have virtually a d.c. problem. With electrolytic conductors this places little restriction on $\omega$, but with liquid metals it is harder to avoid self-inductance effects. For this and other reasons it is more convenient with liquid metals to use a steady field from a d.c. electromagnet or a permanent magnet.

At the low frequencies and non-zero conductivities assumed we may safely neglect displacement current in comparison with the conduction current $\mathbf{j}$. The condition for this is $\omega\epsilon/\sigma \ll 1$, $\epsilon$ being the fluid's permittivity (see Appendix for typical magnitudes). In this book we shall not discuss the use of induction flowmeters with dielectric liquids such as oil where the displacement current is more important than the conduction current. The theory of such devices has been discussed by Cushing.*

In flowmeters with a.c. fields the problem of 'pick-up' also presents itself, since the leads to the electrodes invariably link some of the oscillatory field. The resultant contribution to the output-signal is in quadrature with the useful part induced by the fluid motion and therefore is readily distinguished and eliminated. This problem has been thoroughly treated in the literature (see Bibliography).

Excluding the self-inductance terms $\partial \mathbf{B}/\partial t$ and the displacement current $\partial \mathbf{D}/\partial t$ from Maxwell's equations leads to

$$\operatorname{curl} \mathbf{E} = 0 \tag{2.2}$$

and
$$\operatorname{curl} \mathbf{B} = \mu \mathbf{j}. \tag{2.3}$$

From (2.2) we see that an electric potential $U$ may be defined in the usual way such that
$$\mathbf{E} = -\operatorname{grad} U. \tag{2.4}$$

Equation (2.3) tells us how the induced currents affect the imposed magnetic field and therefore does not concern us until the next chapter. However it does have the consequence that

$$\operatorname{div} \mathbf{j} = 0, \tag{2.5}$$

expressing the fact that local accretion of charge is unimportant at low or zero frequency.

Applying (2.5) to (2.1) with $\sigma$ constant gives the Poisson equation

$$\nabla^2 U = \operatorname{div} \mathbf{v} \times \mathbf{B} \tag{2.6}$$

in view of (2.4) This is the basic flowmeter equation which, together with appropriate boundary conditions, determines the distribution of $U$ from given distributions of $\mathbf{v}$ and $\mathbf{B}$. The same equation has been widely studied by the oceanographers who have also allowed $\sigma$ to vary. The problem is clearly that of finding the potential distribution that results from a space distribution of dipoles, or miniature electric generators, of strength $\mathbf{v} \times \mathbf{B}$ per unit volume. The direction of action of each dipole is perpendicular to both the fluid motion and the magnetic flux, as would be expected. We note that the conductivity $\sigma$ has disappeared from the problem, but it reappears if conditions are specified at conducting boundaries.

Since most fluid motions are turbulent, it is important to consider whether (2.6) still holds in any sense for turbulent flows. If we denote

* Cushing, V. (1958). *Rev. Sci. Instrum.* **29**, 692.

time-mean quantities by a bar and turbulent fluctuations by a dash, (so that $U = \bar{U} + U'$, for instance), taking the time-mean of (2.6) yields

$$\nabla^2 \bar{U} = \mathrm{div}\,(\bar{\mathbf{v}} \times \overline{\mathbf{B}} + \overline{\mathbf{v}' \times \mathbf{B}'}). \qquad (2.7)$$

The second vector product term will be negligible for various reasons: there appears to be no reason to expect any strong correlation between the random velocity and field fluctuations; the quantity $\mathbf{B}'$ is an induced field and, as will be seen in ch. 3, is usually small in comparison with the imposed field even with liquid metals. With electrolytes it is quite negligible. Then, if we leave the last term out of (2.7), it simply becomes (2.6), interpreted so as to apply to the *mean* values of $U$, $\mathbf{v}$ and $\mathbf{B}$. Thus the fact that a flow may be turbulent does not complicate the issue. An electromagnetic flowmeter can still measure the mean velocity.

Using a vector identity we may rewrite (2.6) as

$$\nabla^2 U = \mathbf{B}\,.\,\mathrm{curl}\,\mathbf{v} - \mathbf{v}\,.\,\mathrm{curl}\,\mathbf{B}, \qquad (2.8)$$

in which the last term may be omitted if the magnetic field is not seriously affected by induced currents in the fluid (so that $\mathrm{curl}\,\mathbf{B} = 0$) and in any case if these currents flow perpendicularly to the fluid motion. Then

$$\nabla^2 U = \mathbf{B}\,.\,\mathrm{curl}\,\mathbf{v}. \qquad (2.9)$$

### 2.1.1. *Boundary conditions*

For any given problem to be determined, sufficient boundary conditions must be specified. The second-order equations (2.6) or (2.9) governing the distribution of $U$ require a single electrical boundary condition specified all over the boundary of the conducting region under scrutiny. Within this region there may be interfaces separating liquid and solid conductors over which *two* electrical boundary conditions must be specified. Cases where there is an interface between two liquids or a free liquid surface are of no great practical interest and will not be considered. Part of the boundary of the region of interest will usually consist of hypothetical surfaces where the fluid enters and leaves, which may perhaps be chosen to be sufficiently remote for all electrical disturbances to have fallen to negligible proportions, or at which the relevant boundary condition is obvious from symmetry or other intuitive considerations.

It will frequently be convenient to employ the usual fluid-dynamical boundary condition that $\mathbf{v}$ vanishes at a solid wall at rest. In consequence of this, (2.1) gives

$$\mathbf{j} = -\sigma\,\mathrm{grad}\,U \qquad (2.10)$$

13

for the stationary fluid immediately adjoining the wall. Most of the theoretical results for flowmeters do not in fact depend on this restriction on the velocity distribution. However, we adopt it simply because it is realistic.

I. *Non-conducting walls.* In this case there can be no current flow from the fluid into the wall. Hence $j_n$ and $\partial U/\partial n$ (by 2.10) vanish at the wall, $n$ being measured normally to the wall. The same condition applies when the contact resistance between the fluid and a conducting wall becomes very high.

II. *Conducting walls.* Let the conductivity of the wall material be uniform, with the value $\kappa$, and the contact resistance between the fluid and the wall be $\tau$. For the purposes of analysis we shall assume that $\tau$ is also uniform, although in practice it is well known to be very variable, both in time and space. It is not even always realistic to take contact potential drop as proportional to normal current density, at a given instant and position. However, use of a constant $\tau$ enables us to explore the effect of contact resistance without too much mathematical complication.

Since the wall is at rest, we have the equations

$$\mathbf{j} = -\kappa \operatorname{grad} U \quad \text{and} \quad \nabla^2 U = 0 \qquad (2.11)$$

prevailing in the wall.

At the outer, non-wetted surface of the wall there is usually no normal current flow and $\partial U/\partial n$ vanishes. Even where the wires leading to the voltage-sensing equipment leave the flowmeter we assume no or negligible current flow, as otherwise the output signal of the meter will be depleted by an internal resistance drop. A voltage-measuring device of input impedance high in comparison with the flowmeter impedance between electrodes or a potentiometer must therefore be employed.

At the inner interface between the fluid and the wall there is a potential difference where the normal current traverses the contact resistance. This is expressed by the equation

$$U_f - U_w = \tau \sigma \, \partial U_f/\partial n, \qquad (2.12)$$

$n$ being directed into the fluid from the wall. The suffixes f and w refer to fluid and wall respectively. Continuity of normal current flow also requires that

$$\sigma \partial U_f/\partial n = \kappa \partial U_w/\partial n, \qquad (2.13)$$

in view of (2.10).

These conditions then suffice to define matching solutions to the two equations (2.9) (in the fluid) and (2.11) (in the wall).

14

## 2.1.2. *Flowmeter sensitivity, S*

Before we proceed to solve the flowmeter equations in particular cases, it is here convenient to define a quantity $S$, a dimensionless measure of the performance or the calibration of any induction flowmeter in which the potential difference $U_{XY}$ induced between two electrodes $X$ and $Y$ is used to indicate a flow rate. We define $S$ as

$$S = U_{XY}/LBv_m, \tag{2.14}$$

where $L$ is the length of $XY$, $B$ is a representative value of the imposed flux density and $v_m$ is a mean flow velocity.

In the simple case where the line $XY$, the field, and the motion are all mutually perpendicular, and where both $\mathbf{B}$ and the velocity are rectilinear and uniform, $S$ takes the standard datum value of unity, provided the walls of the pipe bearing the fluid are non-conducting. In this case no induced currents can flow, the fluid polarises and $U_{XY}$ is simply the integral of the constant induced e.m.f. $\mathbf{v} \times \mathbf{B}$ along $XY$. The pipe shape is immaterial.

We shall find that $S$ takes the value unity in certain other cases but is frequently very different from unity.

## 2.2. Two-dimensional induction flowmeter theory

Any real flowmeter will have finite dimensions. The fluid which is being metered must enter and leave the region of imposed magnetic field. The consequences of this entry and exit process will be described as *edge effects* and will be examined later. For the present it will be assumed that the interesting region near the electrodes is sufficiently far removed from the edge regions for the field and flow configurations there to be virtually invariant in the direction of fluid motion. This presupposes that the meter is reasonably long in the flow direction in comparison with the transverse dimensions of the straight pipe bearing the fluid. In addition we shall assume that the mean flow distribution of the fluid changes only relatively slowly in the flow direction, if at all, all velocities being parallel to the pipe axis. In turbulent flow this refers to local mean velocities. The problem is now a two-dimensional one in which all conditions are identical in each cross-section of the pipe, apart from the longitudinal pressure gradient and, in one case to be studied, a longitudinal potential gradient. Boundary conditions now need only to be specified around the two-dimensional cross-section of the pipe.

In view of the fact that a measuring instrument ought to be as

15

absolute as possible, i.e. that its calibration should depend on the smallest number of factors, none of them imponderable, we shall start our survey of flowmeter theory by considering under what conditions a flowmeter's sensitivity will be quite independent of the details of the velocity distribution inside it.

### 2.2.1. *Two-dimensional cases where the value of $S$ is independent of the form of the velocity profile*

I. *Rectangular pipe (transverse-field meter).* Consider the case of a pipe of rectangular cross-section situated in a uniform transverse magnetic field which is parallel to one pair of sides of the pipe. If both these sides of the pipe are made highly conducting in comparison with the fluid and function as electrodes of large area, it is evident that they will exert an 'averaging effect' over the various induced e.m.f.s throughout the fluid. This suggests that the value of $S$ will become independent of the velocity distribution, a fact already noted by various authors.*

Take Cartesian axes as shown in fig. 6, the $x$-axis being parallel to the uniform imposed magnetic field $B$, and the $z$-axis in the direction of the fluid velocity $v$. Let the dimensions of the fluid passage be $2a$ and $2b$ as indicated. The top and bottom walls are taken to be perfectly conducting, each at a uniform potential, the p.d. between them being $U_{XY}$, the flowmeter output signal. The side walls may be either non-conducting or conducting with conductivity $\kappa$ and thickness $w$. There may be contact resistance at the walls. We shall

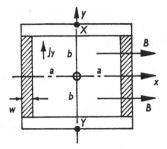

Fig. 6. Rectangular flowmeter with highly conducting walls parallel to the field.

assume that any contact resistance on the perfectly conducting walls is uniform at the value $\tau$. Instead of solving the flowmeter equation (2.9), here we go back to Ohm's Law (2.1) expressed in the $y$-direction,

$$\partial U/\partial y = Bv - j_y/\sigma. \tag{2.15}$$

Integrating this along any line in the fluid parallel to the $y$-axis gives

$$U_{XY} = B\int_{-b}^{b} v\,dy - \frac{1}{\sigma}\int_{-b}^{b} j_y\,dy - \tau\{j_{y(y=b)} + j_{y(y=-b)}\}. \tag{2.16}$$

* Arnold, J. S. (1951). *Rev. Sci. Instrum.* 22, 43. Shercliff, J. A. (1952). *A.E.R.E. (Harwell) Report* X/R 1052. Thürlemann, B. (1955). *Helv. Phys. Acta,* 28, 483. Holdaway, H. W. (1957). *Helv. Phys. Acta,* 30, 85.

The last term represents the effect of contact resistance on the current flowing normally at the top and bottom walls. If the side walls are conducting there will be a similar equation

$$U_{XY} = -\frac{1}{\kappa} \int_{-b}^{b} j_{yw} dy \qquad (2.17)$$

in which $j_{yw}$ is the $y$-component of current in the walls. The joint between the walls has been assumed to present no contact resistance.

Since the voltage-measuring device is assumed to draw negligible current from the electrodes and there is no axial current flow, the net current flow across any line parallel to the $x$-axis must be zero. This leads to the equation

$$\int_{-a}^{a} j_y dx + \int j_{yw} dx = 0, \qquad (2.18)$$

the second integral being taken through both side walls. Integrating and combining (2.16), (2.17) and (2.18) gives

$$U_{XY}(2a\sigma + 2w\kappa) = \sigma B \int_{-b}^{b} \int_{-a}^{a} v\, dx\, dy - \tau\sigma \int_{-a}^{a} (j_{y(y=b)} + j_{(y=-b)})\, dx \qquad (2.19)$$

in which the first term on the right-hand side may be replaced by $4ab\sigma Bv_{\mathrm{m}}$. The value of $S$ may be deduced from (2.19) in a form independent of the velocity distribution in a number of cases:

(a) *Non-conducting side walls* ($\kappa = 0$). In this case $j_{yw}$ and hence $\int_{-a}^{a} j_y dx$ vanish. The last term disappears from (2.19) and we deduce

$$S = U_{XY}/2bBv_{\mathrm{m}} = 1.$$

We see that the highly conducting walls do indeed exert their averaging effect and the signal is the same as if the velocity was uniform, even in the presence of uniform contact resistance.

(b) *No contact resistance on the highly conducting walls* ($\tau = 0$). From (2.19) we now get

$$S = \frac{1}{1+d}, \qquad (2.20)$$

where $d$ stands for the dimensionless expression $w\kappa/a\sigma$, which measures the importance of the short-circuiting effect of the conducting walls. Significantly $S$ is now less than unity. Contact resistance, which may be non-uniform, on the side walls does not affect this result. The equation (2.20) will be a good approximation even when the top and bottom walls are not highly conducting, provided $b/a$ is reasonably large. Then variations in potential along these walls are small in comparison with $U_{XY}$. Values of $b/a$ larger than 5 will make (2.20) accurate enough for most purposes in this case.

(c) *Very high contact resistance on the side walls.* This case might come about if it was essential to have an all-metal pipe, but it was possible to coat the side walls with some insulator. In this case there is no current flow between the fluid and the side walls and $\int j_{yw} dx$ is constant and equal to $-2U_{XY}\kappa w/2b$. In view of (2.18), the last term of (2.19) now becomes

$$2\tau\sigma U_{XY}\kappa w/b \tag{2.21}$$

and hence

$$S = \frac{1}{1+d(1+\sigma\tau/b)}. \tag{2.22}$$

The extra term in the denominator represents the loss of signal due to the contact resistance on the highly conducting walls associated

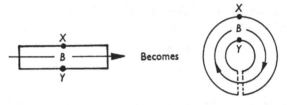

Fig. 7

with the short-circuit currents in the side walls. When the side contact resistance is not high, (2.22) will give a lower or pessimistic limit to the value of $S$, since $\int j_{yw} dx$ will be least when $y = \pm b$, for all reasonable velocity profiles, and so the last term in (2.19) is less than is given by (2.21).

It is apparent that the rectangular flowmeter has desirable features if a calibration which is insensitive to velocity profile is required, particularly if the aspect ratio $b/a$ can be made large. This configuration has the further attraction that it is easy to make a magnet to provide a uniform field across a narrow rectangular region.

II. *Circular pipe and circular field (axial-current meter).* This case is a generalisation of case (a) above. If we imagine the rectangular channel distorted by being bent in cross-section, as suggested in fig. 7, until the two side walls coincide and can be omitted, the resultant channel is now an annulus. The field lines must be simultaneously bent into circles concentric with the annulus so as to remain parallel to the walls that function as electrodes. Such a field can be provided if there is an imposed, concentric axial flow of current, either in the fluid or in a solid conductor within the annulus. The inner wall of the annulus may even be omitted, leaving simply a circular pipe with a point electrode on the centre-line. Figure 3 illustrated the case of a circular pipe with the axial current flowing in the fluid, while fig. 8 shows an annular meter in which the axial current flows in a central solid conductor. The fluid motion and the magnetic field are still perpendicular to one another and now the induced e.m.f. acts radially.

If the pipe walls are highly conducting in comparison with the fluid and function as electrodes, we should expect them to exert the desirable averaging effect over the various e.m.f.s induced by the fluid motion, just as occurs in the closely related rectangular meters

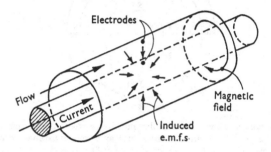

Fig. 8. Annular flowmeter with axial current in central conductor.

previously discussed. Note that it is not merely necessary that the wall conductivity be high. Obviously the walls must not be too thin also. The precise condition is that $d(= w\kappa/a\sigma)$ should be large (where $w$ = wall thickness, $\kappa$ = wall conductivity, $a$ = pipe radius or annular gap, $\sigma$ = fluid conductivity). The presence of high *uniform* contact resistance will assist the walls in exerting their averaging effect, as is evident from the equivalent circuit in fig. 9, provided that negligible net current is being drawn from the region either by the external voltage-measuring circuit or by short circuits at the ends of the meter, as is discussed later in §2.3.

The variation of the flux density $B$ with distance $r$ from the axis of the pipe has still to be specified. If $B$ is directly proportional to $r$ an exact averaging effect is obtained, and $S$ is insensitive to the velocity distribution. This is formally proved later. The reason for this result is that the flow through an annular element of cross-section at radius $r$ is obviously proportional to $r$. Therefore having $B$ proportional to $r$ correctly weights the contributions at different radii to the total induced p.d. This type of field results from an axial energizing current *in the fluid* of uniform density $j$. This is the case, mentioned earlier,

2-2

where there is inevitably a potential gradient along the pipe. Equation (2.3) then implies that

$$B = \tfrac{1}{2}\mu r j. \qquad (2.23)$$

For this result to apply exactly in the annular case also, in addition to the current $j$ in the fluid, a current must be supplied concentrically but not necessarily uniformly in a central conductor so as to give a mean current density of $j$ over the circle within the annulus. In

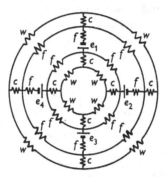

Fig. 9. Equivalent circuit for annular flowmeter with conducting walls. The resistors $c$, $f$ and $w$ represent contact, fluid and wall resistance, respectively, while $e_1$, $e_2$, etc., represent the non-uniform e.m.f.s induced by the non-uniform velocity. Increasing all the $c$ resistors or lowering all the $w$ resistors tends to make the p.d. between the two 'walls' approach $\tfrac{1}{4}(e_1 + e_2 + e_3 + e_4)$ everywhere.

practice these complications may be avoided by having an axial current flowing only in the central conductor and not in the fluid. Then $B$ in the fluid varies *inversely* with $r$, but provided the inner radius of the annulus is not too small, the voltage induced between the two walls is not particularly sensitive to the velocity profile, and an adequate averaging effect is achieved. This point is investigated at the end of this section.

Suppressing the axial current in the fluid has several points in its favour. One is that the highly conducting pipe walls would carry an excessive and rather indeterminate fraction of the total current supplied to the meter. This would waste power and render uncertain the calibration of the meter, which depends on the current *in the fluid*. The alternative way of avoiding this difficulty would be to install laminated walls giving high circumferential but zero axial conductivity. This would be a worthwhile complication if a flowmeter having absolute insensitivity to velocity profile was essential.

Another important point is that suppressing the axial potential gradient in the fluid means that the relative axial positioning of the electrodes is no longer crucial. In any case it is more economical to put the axial current through a central copper conductor than through a liquid metal of inferior conductivity, and it is essential if the fluid is an electrolytic conductor.

We shall now formally demonstrate that the circular axial-current meter with $B$ proportional to $r$ does indeed give a sensitivity wholly independent of the velocity distribution provided wall conductivity is very high, even when there is *uniform* contact resistance $\tau$ at the wall. The fluid conductivity $\sigma$ may also be permitted to vary with radius, as might occur in the case of an axisymmetric temperature profile.

In view of (2.23), the radial component of Ohm's Law becomes

$$j_r/\sigma = -\partial U/\partial r + \tfrac{1}{2}\mu jrv.$$

We are using polar co-ordinates, concentric with the pipe. Integrating over the cross-section of the annulus between the inner and outer radii, $b$ and $a$ respectively, gives

$$\int_b^a \frac{dr}{\sigma} \int_0^{2\pi} j_r\, d\theta = -\int_0^{2\pi} (U_a - U_b)\, d\theta + \tfrac{1}{2}\mu j \int_0^{2\pi}\int_b^a vr\, dr\, d\theta. \quad (2.24)$$

Now $\displaystyle\int_0^{2\pi} j_r\, d\theta$ vanishes since there is no net current flow across concentric cylindrical surfaces in the fluid, divergence of the axial current flow being zero. The output signal of the meter, $U_{XY}$, is given by

$$U_{XY} = U_a - U_b - (\tau j_r)_a + (\tau j_r)_b.$$

The last two terms represent the contact resistance drops. Integrating this equation gives

$$2\pi U_{XY} = \int_0^{2\pi} (U_a - U_b)\, d\theta,$$

since $\tau$ and $U_{XY}$ are independent of $\theta$. From (2.24) it follows that

$$U_{XY} = \mu j Q/4\pi, \quad (2.25)$$

in which $Q$ is the volumetric flow rate $\displaystyle\int_0^{2\pi}\int_b^a vr\, dr\, d\theta$. This may be put into the familiar standard form

$$S = U_{XY}/LB_m v_m = 1,$$

where $L$ is taken equal to the angular gap $(a-b)$, $B_m$ is evaluated at the mean radius $\tfrac{1}{2}(a+b)$ and $v_m$ is the mean velocity, $Q/\pi(a^2-b^2)$. It should be noted that the case of the circular pipe with a point electrode at the centre is included as the case $b = 0$ in the above analysis.

A final question to investigate before passing on is how far the output signal becomes sensitive to the radial distribution of velocity if the field $B$ does not vary linearly with $r$, as in meters with the axial current flowing solely in a central conductor. The highly conducting walls would still exert a strong averaging effect over variations of velocity and induced e.m.f. at different values of $\theta$. To explore the effect of velocity variation with $r$, consider two extreme velocity profiles, both independent of $\theta$, one a uniform velocity and the other a parabolic distribution.

If there is no axial current in the fluid the field $B$ now equals $\mu J/2\pi r$, where $J$ is the total axial current. If the velocity is uniform, we find that

$$U_{XY} = \mu J Q \log (a/b)/2\pi^2(a^2-b^2), \qquad (2.26)$$

whereas if the velocity varies quadratically with $r$ and is zero at the inner and outer walls, so that

$$v = 6Q(a-r)(r-b)/\pi(a^2-b^2)(a-b)^2,$$

then
$$U'_{XY} = 3\mu J Q\{\tfrac{1}{2}(a^2-b^2) - ab \log (a/b)\}/\pi^2(a^2-b^2)(a-b)^2.$$

The ratio $U'_{XY}/U_{XY}$ is equal to

$$3\{(\gamma+1)/(\gamma-1)\log \gamma - 2\gamma/(\gamma-1)^2\}, \quad \text{if} \quad \gamma = a/b,$$

or approximately $1-(\gamma-1)^2/30$ when $\gamma$ is near unity. Even for $\gamma$ as large as 5 the ratio has only fallen to 0·92 whereas for the more likely practical value of 2, the difference between the two values of $U_{XY}$ is only 2 per cent. The conclusion is that the dependence of the sensitivity on the velocity profile is still very weak even when $B$ ceases to vary linearly with $r$, and (2.26) is good enough for most purposes. This is equivalent to the statement

$$S \equiv U_{XY}/LB_{\mathrm{m}}v_{\mathrm{m}} = 1,$$

where $L = (a-b)$, $v_{\mathrm{m}} = Q/\pi(a^2-b^2)$ and $B_{\mathrm{m}}$ is the mean flux density $\mu J \log (a/b)/2\pi(a-b)$.

2.2.2. *Two-dimensional, circular pipe cases where the value of $S$ is independent of the form of the velocity profile, provided this is axisymmetric*

The requirement that the velocity profile in a circular pipe be axisymmetric is not unreasonable. It is common practice to mount fluid meters of all types well downstream of flow disturbances such as bends and cocks. The induction flowmeter is at most no more demanding in this respect than its competitors, and it may be much better. With liquid metals flowing in transverse-field meters, magnetic forces may result in the disruption of the natural axisymmetry of pipe flow if the conductivity and field strength are high enough. This effect is studied in the next chapter. In this section we shall assume that the velocity profile has not departed from axisymmetry.

I. *Axial-current meter*. Here we require an axial energizing current (of density $j$) in the fluid. If the flow channel is an annulus, there must also be an axial current of appropriate density in a central conductor. Then the circular field is given by

$$B = \tfrac{1}{2}\mu jr, \tag{2.23}$$

and it is plain that, if the velocity is a function of $r$ only, the radially induced e.m.f.s will result in no radial current flow, whether there is wall conductivity and contact resistance, which may be non-uniform, or not. None of the induced e.m.f. is lost in resistance drops and therefore

$$dU/dr = \tfrac{1}{2}\mu jrv$$

and $U_{XY} = \tfrac{1}{2}\mu j \int rv\,dr = \mu jQ/4\pi$ again, as in (2.25). The difference from the earlier case is that here the walls are not highly conducting and may even be non-conducting, since no $\theta$-wise averaging effect is called for when the velocity profile is axisymmetric. If the walls are non-conducting the electrodes must have direct access to the liquid. Their circumferential position is immaterial, but they must lie in exactly the same cross-section of the pipe because of the axial voltage gradient.

We have already seen how a different relationship between $B$ and $r$ causes this kind of flowmeter to lose its property of exactly integrating the velocity profile to yield the total flow, although the consequent errors may be small.

II. *Transverse-field meter*. We now consider meters of the Williams type, illustrated in fig. 10 and consisting of a circular pipe under a uniform transverse field with two electrodes mounted at the ends of a diameter that is perpendicular both to the field and the flow. The pipe walls may be conducting and there may be uniform contact resistance. When conducting walls are used it is common practice to use external electrodes $A$ and $D$ as shown in fig. 11. Otherwise, the internal electrodes $B$ and $C$ must be employed. The electrodes are now point electrodes (or, perhaps, line electrodes extending some distance axially). In contrast, in the flowmeters discussed in §2.2.1, the electrodes were surfaces of finite area, except when there was axial current

23

flow in the fluid, in which case the electrodes were points (or perhaps, line electrodes extending round the periphery of the pipe).

If the velocity profile is axisymmetric and if the conductivities of the fluid and the wall and the contact resistance are all constant, we shall find that the sensitivity of the meter is independent of the velocity distribution. This remarkable result occurs despite the fact that the extent to which some of the induced e.m.f. is lost in

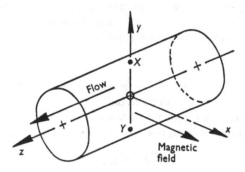

Fig. 10. Circular, transverse-field flowmeter.

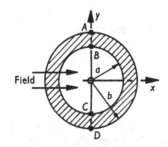

Fig. 11. Cross-section of transverse-field meter, showing electrode positions and axes.

driving circulating currents depends on the velocity distribution. For instance a parabolic velocity distribution in a non-conducting pipe permits a quarter of the e.m.f. induced along the line $BC$ (in fig. 11) to be lost Ohmically, whereas for a uniform velocity distribution in the same pipe none of the induced e.m.f. is lost. Yet both these cases give the same signal between $B$ and $C$ for a given total flow rate or mean velocity.

We take Cartesian axis such that $Ox$ is parallel to the uniform imposed field $B$ and $z$ is parallel to the direction of the flow velocity $v$.

24

If the suffices f and w refer to fluid and wall quantities, the potential $U$ is governed by the equations

$$\nabla^2 U_{\text{w}} = 0, \quad \nabla^2 U_{\text{f}} = B \partial v/\partial y, \qquad (2.27)$$

from (2.11) and (2.9) respectively. The Laplacians comprise two derivatives only. Equation (2.9) applies here even without the approximation curl $\mathbf{B} = 0$ because the induced currents flow in the $(x, y)$-planes, perpendicular to the velocity, so that $\mathbf{v} . \text{curl} \, \mathbf{B} = 0$. Since we are assuming $v = v(r)$, the second of equations (2.27) becomes

$$\nabla^2 U_{\text{f}} = Bv' \sin \theta, \qquad (2.28)$$

where a dash denotes differentiation with respect to $r$. Taking the internal and external radii of the pipe wall as $a$ and $b$ respectively, we have the following boundary conditions:
when $r = a$
$$U_{\text{f}} + \sigma \tau \partial U_{\text{f}}/\partial r = U_{\text{w}} \quad \text{(from (2.12))},$$
and $\qquad \sigma \partial U_{\text{f}}/\partial r = \kappa \partial U_{\text{w}}/\partial r \quad \text{(from (2.13))};$ $\qquad (2.29)$
when $r = b$ $\qquad \partial U_{\text{w}}/\partial r = 0.$

The solutions
$$U_{\text{f}} = Z(r) \sin \theta$$
and $\qquad U_{\text{w}} = (Ar + C/r) \sin \theta, \quad (A \text{ and } C \text{ const.}),$

satisfy (2.27), (2.28) if

$$r^2 Z'' + r Z' - Z \equiv (\partial/\partial r)(r^2 Z' - rZ) = r^2 Bv', \qquad (2.30)$$

and the boundary conditions (2.29) if

$$Aa + C/a = Z(a) + \sigma \tau Z'(a),$$
$$\kappa (A - C/a^2) = \sigma Z'(a) \quad \text{and} \quad A = C/b^2. \qquad (2.31)$$

Integrating (2.30) (the right-hand side by parts) from $r = 0$ to $a$, gives

$$a^2 Z'(a) - a Z(a) = B[r^2 v]_0^a - B \int_0^a 2rv \, dr = -Ba^2 v_{\text{m}}, \qquad (2.32)$$

since $v$ vanishes at the wall where $r = a$. Eliminating $Z(a)$, $Z'(a)$, $A$ and $C$ from (2.31) and (2.32) yields the sensitivity

$$S = \frac{U_{AD}}{2bBv_{\text{m}}} = \frac{2a^2}{(a^2 + b^2) + (\kappa/\sigma)(1 + \sigma \tau/a)(b^2 - a^2)}. \qquad (2.33)$$

The length in the denominator of $S$ has been taken as the distance $2b$

25

between the exterior electrodes $A$ and $D$. Putting $b = a$ in (2.33) yields the result

$$S = \frac{U_{BC}}{2aBv_{\mathrm{m}}} = 1, \qquad (2.34)$$

appropriate to the case of non-conducting walls. This result was derived by Williams for certain particular velocity distributions and he guessed it might be true more generally. The general result (2.34) was first proved by Thürlemann* and a shorter proof due to W. E. Lamb was given by Kolin.† The proof used here is due to J. W. Poduska.

The result (2.34) can also be applied to the case of conducting walls or concentric solid deposits that have the same conductivity as the fluid, since these behave like a stagnant, axisymmetric layer of fluid. There must not be any contact resistance, however. This is borne out if we put $\tau = 0$ and $\kappa = \sigma$ in (2.33), with the result that

$$S = \frac{U_{AD}}{2bBv_{\mathrm{m}}} = \frac{U_{AD}\pi a^2}{2bBQ} = \frac{a^2}{b^2}. \qquad (2.35)$$

Here the mean velocity $v_{\mathrm{m}}$ is based on an area $\pi a^2$, whereas to apply (2.34) to this case, $v_{\mathrm{m}}$ should be based on the whole conducting area $\pi b^2$. This reconciles (2.34) and (2.35). Thürlemann* and Kolin‡ have both treated this case. Kolin points out that it corresponds closely to the case of blood flow in arteries, and that, as (2.35) shows, $U_{AD}$ depends only on $B$, $b$ and the flow rate $Q$. Thus, for given $b$, the calibration of the meter does not depend on the thickness of the wall of the artery, which may be difficult to ascertain.

Elrod & Fouse§ developed the result (2.33) for the case $\tau = 0$, and Astley‖ found a result very similar to (2.33) by taking a layer of stagnant, poorly conducting fluid of finite thickness at the walls.

The result (2.33) simplifies if the wall thickness $w$ is small in comparison with $a$. Then, if we put $d = w\kappa/a\sigma$,

$$S = \frac{U_{AD}}{2aBv_{\mathrm{m}}} = \frac{1}{1 + d(1 + \sigma\tau/a)}. \qquad (2.36)$$

This result follows directly from (2.33) if we let $a \to b$, keeping $\kappa(b-a)$ finite, or may be deduced by solving (2.28) subject to the special

* Thürlemann, B. (1940). *Helv. Phys. Acta,* **13**, 343 and (1941), **14**, 383.

† Kolin, A. (1945). *Rev. Sci. Instrum.* **16**, 109.

‡ Kolin, A. (1952). *Rev. Sci. Instrum.* **23**, 235.

§ Elrod, H. G. & Fouse, R. R. (1952). *Trans. Amer. Soc. Mech. Engrs,* **74**, 589. See also: Gessner, U. (1961). *Biophys. J.* **1**, 627.

‖ Astley, E. R. (1952). *General Electric Report* R 52 GL 42. See also: Gray, W. C. & Astley, E. R. (1954). *Instrum. Soc. Amer. J.* **1**, 15.

boundary conditions appropriate to thin walls, developed in the next section. Thürlemann* arrived at the result (2.36) in the case $\tau = 0$, by arguing that the walls could be imagined to be equivalent to walls of thickness $\kappa w/\sigma$ and of the same conductivity $\sigma$ as the fluid. He did not state the necessary restriction that the walls be thin, however. Then he used his analysis appropriate to the case $\kappa = \sigma$, mentioned earlier.

Before leaving these various results we remark again that in each case the value of $S$ is independent of the velocity profile if this is axisymmetric.

### 2.2.3. *Thin-wall boundary conditions*

Figure 12 shows a portion of such a wall, the flow being perpendicular to the page and all currents flowing in the plane of the page. Let $J$ denote the total current in the wall at any point per unit length of pipe. Ohm's Law in the wall gives

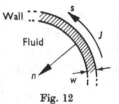

Fig. 12

$$J = -w\kappa\,\partial U_{\text{w}}/\partial s, \qquad (2.37)$$

$s$ being measured along the wall. $J$ varies in accordance with the equation

$$\partial J/\partial s = -j_{\text{n}} = \sigma\,\partial U_{\text{f}}/\partial n \quad \text{(by (2.10))}. \qquad (2.38)$$

The equations (2.13), (2.37) and (2·38) yield the condition that $U_{\text{f}}$ must satisfy

$$\sigma\frac{\partial U_{\text{f}}}{\partial n} + w\kappa\frac{\partial^2}{\partial s^2}\left(U_{\text{f}} - \tau\sigma\frac{\partial U_{\text{f}}}{\partial n}\right) = 0, \qquad (2.39)$$

if we take $w\kappa$ as constant. In polar co-ordinates $n$ is replaced by $-r$, $s$ by $r\theta$.

### 2.2.4. *Two-dimensional analysis of the transverse-field, circular flowmeter for a general velocity profile*

Having now seen how the transverse-field circular meter is unaffected by symmetrical variations of the velocity profile, we turn next to the question as to how severely asymmetry of the profile does alter the sensitivity. Asymmetry could arise from upstream disturbances or, in the case of liquid metals, from the action of magnetic forces.

---

* Thürlemann, B. (1941). *Helv. Phys. Acta*, **14**, 383.

27

To render the analysis reasonably simple, this section will be restricted to the case of non-conducting walls. Again the equation to be solved is

$$\nabla^2 U = B\partial v/\partial y, \qquad (2.40)$$

subject to the boundary conditions $v = 0$ and $\partial U/\partial r = 0$ at the wall where $r = a$. The problem being linear, it is possible to find a superposition solution which gives $U_{BC}$ (electrodes $B$ and $C$ as in fig. 11) as an integral, over the pipe cross-section, of the local velocity $v$ times a weight function, which indicates the ability of flow at various points to contribute to $U_{BC}$. The solution of (2.40), giving $U$ as a function of position $(x, y)$, is

$$U = \mathscr{I}\left\{\frac{B}{2\pi}\int\int v(\xi, \eta)\left[\frac{1}{\zeta-z}+\frac{a^2}{\overline{\zeta}(a^2-z\overline{\zeta})}\right]d\xi\,d\eta\right\}, \qquad (2.41)$$

in which $\mathscr{I}$ denotes 'imaginary part of', and $\xi$ and $\eta$ are Cartesian co-ordinates referred to the $(x, y)$ axes, $\zeta = \xi + i\eta$, $\overline{\zeta} = \xi - i\eta$ and $z = x + iy$. The double integral is taken over the circular cross-section of the pipe. Setting $z = \pm ia$ in (2.41) leads to the result

$$S = \frac{U_{BC}}{2aBv_{\mathrm{m}}} = \frac{\int\int v(x, y)\,W(x, y)\,dx\,dy}{\int\int v(x, y)\,dx\,dy}, \qquad (2.42)$$

in which $W$, the weight function, is given by

$$W = \frac{a^4+a^2(x^2-y^2)}{a^4+2a^2(x^2-y^2)+(x^2+y^2)^2} = \frac{a^4+a^2r^2\cos 2\theta}{a^4+2a^2r^2\cos 2\theta+r^4}.$$

Figure 13 shows the contours of $W$. The minimum value of $W$ within the pipe is $0.5$ (at the side walls) and $W$ increases without limit near the electrodes $B$ and $C$. This result permits one to see at a glance the effect of flow at various parts of the pipe on the output signal. For instance, if $v$ is everywhere positive and there is no reverse flow, then, since $W \geqslant \frac{1}{2}$, $\int\int vW\,dx\,dy \geqslant \frac{1}{2}\int\int v\,dx\,dy$ and $S \geqslant \frac{1}{2}$ also. For $S$ to reach the value $1/2$ without reverse flow, the motion would have to be concentrated near the side walls. Should the flow be concentrated near one or both electrodes, however, $S$ can increase without limit. Close downstream of an obstruction it is quite likely that eddies involving reverse flow could occur. Moreover, if there were a small amount of reverse flow near an electrode, where $W$ is large, this might produce a disproportionate reduction in $S$, counteracting the effect of the larger forward flow in other parts of the pipe, with the result that $S$ could fall below $1/2$ and even become negative. These phenomena have been observed experimentally.*

* Shercliff, J. A. (1955). *J. Sci. Instrum.* **32**, 441.

It is clear that the circular, transverse-field meter is far from being the absolute instrument that it has so often erroneously been claimed to be. Few other fluid meters are liable to errors larger than $\pm 100$ per cent in the presence of upstream disturbances! But this is to take too pessimistic a view. If a reasonably long settling length is inserted between any violent flow disturbance and a circular, transverse-field

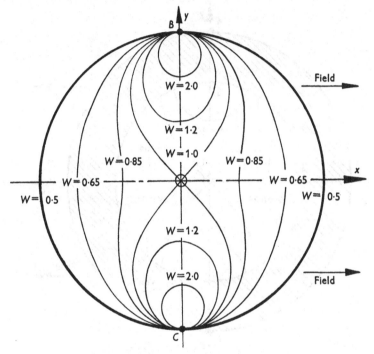

Fig. 13. The quantity $W$ indicates the ability of flow at various parts of the cross-section to contribute to the output signal.

flowmeter, then the velocity profile will normally be sufficiently symmetrical at the meter for the results of §2.2.2 to apply. What exactly constitutes 'reasonably long' is still somewhat uncertain, and could only be established by a programme of systematic tests. It seems that induction flowmeters are less demanding in this respect than orifice or venturi meters, and if a type which is absolutely independent of velocity profile is selected there need be no restriction on upstream conditions.

Head's experience* is that even circular transverse-field meters

* Head, V. P. (1959). *Trans. Amer. Soc. Mech. Engrs*, **81**, 660.

are not much affected by upsteam conditions in practice. Other tests* revealed deviations in sensitivity up to 3 per cent in a circular transverse-field meter when a partly opened gate-valve was placed approximately four diameters upstream of the electrode plane. Probably the reason the deviations were not larger is that such an obstruction produces violent turbulence over the whole pipe rather

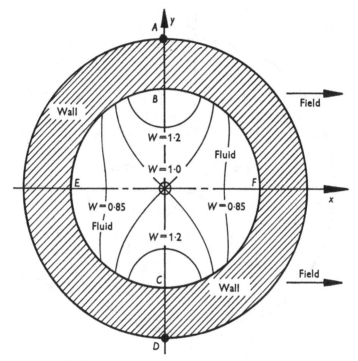

Fig. 14. $W$-distribution in the case of a pipe with walls of the same conductivity as the fluid.

than well-defined areas of slow and fast flow. The same series of tests* showed that swirl had no detectable effect on the performance of circular flowmeters.

There remains the question how these conclusions would be modified if the circular transverse-field meter had conducting walls, possibly with contact resistance. The general results would be similar but the effects more moderate. Flow near $B$ would no longer have such a strong effect on an output signal measured between external

* Worcester (Mass.) Polytechnic, Alden Hydraulics Lab. (April 1955). *Report on Foxboro meter.*

electrodes $A$ and $D$ (see fig. 14). In the case where the conductivities of the fluid and walls are the same and there is no contact resistance, the distribution of $W$ shown in fig. 13 can still apply. Consider a case where the wall thickness is half the pipe's inner radius. Figure 14 reproduces fig. 13 with the fluid and wall regions for this case indicated. Only the values of $W$ inside the fluid region are relevant now. The extreme values of $W$ in this case are then 1·8 at $B$ and $C$ and 0·7 at $E$ and $F$ (in comparison with the previous limits of $\infty$ and 0·5) and the likelihood that $S$ will vary seriously is correspondingly diminished. Note that use of these values of $W$ leads to a figure for $S$ based on a mean velocity taken over the circular area which encloses the fluid *and pipe walls*. The value of $S$ based on the true mean velocity of the fluid (taken over the circular area of fluid only) would be lower in the ratio (pipe I.D./pipe O.D.)$^2$.

Uniform contact resistance would have the further effect of blurring out the influence of local flow variations.

### 2.2.5. *Rectangular, transverse-field flowmeters*

Rectangular meters are not normally used, simply because circular piping is so much more convenient, unless a special effort is being made to secure a flowmeter sensitivity that is independent of velocity distribution, either by having a rectangular channel that is narrow in the field direction or by the use of highly conducting walls parallel to the field. For this reason we shall not carry out a general analysis of the dependence of a rectangular flowmeter's output on the velocity distribution in other cases. Just as in circular pipes there will be considerable sensitivity to the form of the velocity profile in general.

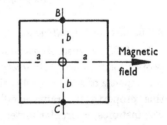

Fig. 15. Cross-section of rectangular flowmeter.

Consider the case of a rectangular flowmeter with non-conducting walls and electrodes $B$ and $C$, as shown in fig. 15. If the velocity profile is uniform, no currents will flow and the sensitivity will take the standard value

$$S = U_{BC}/2bBv_{\mathrm{m}} = 1.$$

On the other hand, a fully developed laminar velocity profile will be associated with circulating currents and the value of $S$ will be different. The equation (2.40) may easily be solved, using the standard velocity profile for laminar flow in rectangular pipes. The resultant values of $S$ for various values of the aspect ratio $a/b$ are given in fig. 16. As would be expected, $S$ tends to unity as $a$ becomes small in comparison with $b$, but, at $a/b = 2.5$, $S$ is larger by 21 per cent, a severe variation. Turbulence would make the velocity more nearly uniform and $S$ would be nearer to unity. In ch. 3 it will be seen that magnetic forces have the same tendency. More severely distorted or asymmetric flows could lead to much greater variations of $S$, particularly if high velocities occurred near the electrodes.

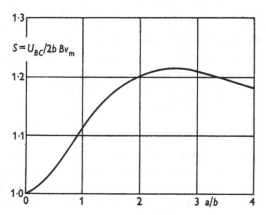

Fig. 16. $S$ for rectangular flowmeters with non-conducting walls as a function of aspect ratio $a/b$.

### 2.2.6. *Integrated-voltage flowmeters*

In §2.2.1 we discussed special types of flowmeters that gave signals proportional to flow rate irrespective of the velocity distribution. At the expense of some complication it is possible in principle to extend this property to transverse-field flowmeters of *any* cross-section. For instance, a circular meter may be made to have invariant sensitivity even when the velocity profile departs from axisymmetry.

Consider a flowmeter of any cross-section. Figure 17 gives an example. Taking first the case of non-conducting walls, we may integrate the equation $j_y/\sigma = -\partial U/\partial y + Bv$ over the entire cross-section, noting that

$$\int_P^Q j_y \, dx = 0, \quad \int (\partial U/\partial y) \, dy = U_L - U_M \quad \text{and} \quad \iint v \, dx \, dy = Q,$$

the volumetric flow rate. Hence

$$\oint U \, dx = BQ,$$

32

where $\oint$ denotes an integral taken round the periphery in a clockwise direction, a result due to Smyth.* Thus by recording the voltage at a sufficient number of points round the periphery and taking an appropriately weighted sum (by the use of passive networks of impedance high compared with the fluid, or otherwise) so as to approximate to $\oint U dx$, it is possible to emerge with a value for $Q$ that is independent of the velocity distribution. This result stands in close relation to 2.2.1 (I). When the pipe is rectangular, $\oint U dx$ simply becomes $(\int U dx)_{\text{top}} - (\int U dx)_{\text{bottom}}$. The use of highly conducting walls to extract an average value of $U$ at the top and bottom walls is seen to be an obvious stratagem. Failing conducting walls, some other way of finding $\int U dx$ or $U_{\text{mean}}$ at the top and bottom walls is called for.

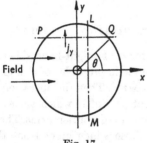

Fig. 17

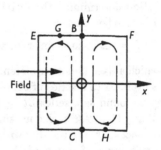

Fig. 18. Rectangular flowmeter; alternative electrode positions.

For the most likely velocity profiles, where there is a steady fall in velocity from the middle of the pipe to the walls, the current circulation will be as shown in fig. 18. The voltage at the top wall will be a maximum at the centre $B$ and will fall towards the corners $E$ and $F$. The mean voltage will prevail at some intermediate point such as $G$, the quarter-point. This suggests that electrodes placed at $G$ and $H$, say, will lead to less variation of $S$ than if they were at $B$ and $C$, for most likely velocity profiles, laminar, turbulent or magnetically distorted. It is now clearer why, for laminar flow, $S$ based on $U_{BC}$ rises significantly above unity as fig. 16 showed.

In the case of a circular channel of radius $a$, illustrated in fig. 17, the result becomes $a \oint U \sin\theta \, d\theta = BQ = B\pi a^2 v_{\text{m}}$, where the integral is now taken anticlockwise round the periphery. One way of realising this as a measuring technique would be to scan the voltage $\Delta U$ between the ends of a diameter rotating at a steady value of $d\theta/dt$

* Smyth, C. (1961). S.B. Thesis, Mech. Eng. Dept., M.I.T.

while varying $B$ proportionally to $\sin \theta$ in synchronism, and to record the mean voltage. Then
$$\pi a \Delta U_{\text{mean}} = B_{\text{max}} Q.$$

Alternatively one of the electrode positions may be kept fixed since the mean voltage at a fixed point is now zero, referred to the same datum as for the moving electrode. Then $2\pi a \Delta U_{\text{mean}} = B_{\text{max}} Q$. We are here assuming the frequency is low enough for skin-effect to be negligible.

There is another instructive way of looking at these results for circular pipes. Let the velocity profile be analysed into components of the form $v_n(r) \sin$ or $\cos n\theta$, where the case $n = 0$ is the axisymmetric component, which carries *all* the net flow $Q$, i.e. $\iint v_0 \, dx \, dy = Q$. The equation determining the distribution of $U$ is the second of (2.27), which here becomes

$$\nabla^2 U = B\Sigma\{v'_n \sin\theta(\sin \text{ or } \cos n\theta) + n(v_n/r)\cos\theta(\cos \text{ or } - \sin n\theta)\},$$

in which a dash denotes differentiation for $r$. The right-hand side can be simplified to a series of sinusoidal terms in multiples of $\theta$, and $U$ may be found as a series $\Sigma U_m(r) \sin$ or $\cos m\theta$. Then the operation of finding $\oint U \sin\theta \, d\theta$ picks out simply that part of $U$ which varies as $\sin\theta$, because of the well-known orthogonality of Fourier terms. There are two sorts of terms of this type, those which stem from the axisymmetric part of the velocity profile (just as in §2.2.2 (II)) and those which stem from the velocity component

$$v_2(r) \cos 2\theta.$$

That part of $U$ of the form $Z(r) \sin\theta$ which arises from this is given by the equation $\nabla^2(Z \sin\theta) = B\{v'_2(-\tfrac{1}{2}\sin\theta) + 2(v_2/r)(-\tfrac{1}{2}\sin\theta)\}$. The $\sin 3\theta$ terms which arise from this velocity component do not concern us. When $r = a$ both $Z'$ and $v_2$ must vanish. The equation for $Z$ is

$$Z'' + Z'/r - Z/r^2 = -B(\tfrac{1}{2}v'_2 + v_2/r)$$

or
$$\frac{d}{dr}(r^2 Z' - rZ) = -(\tfrac{1}{2}r^2 v'_2 + r v_2).$$

Integrating from 0 to $a$ gives

$$a^2 Z'(a) - a Z(a) = -[\tfrac{1}{2}r^2 v_2]_0^a = 0,$$

and $Z(a)$ therefore vanishes, i.e. this velocity component does not contribute to $\oint U \sin\theta \, d\theta$ round the periphery. Thus this integral singles out the contribution of the axisymmetric part of the profile and we already know that an axisymmetric profile leads to a sensitivity independent of the velocity variation. It is now clearer why taking the integral $\oint U \sin\theta\, d\theta$ leads to invariant sensitivity because all but the axisymmetric part of the profile is thereby ignored.

So far no fresh information has emerged, but the way is now pointed to extending the result to circular pipes with conducting walls, where once again the axisymmetric part of the velocity profile gives rise to a contribution to $U$ at the outer periphery that is proportional to $\sin\theta$ and may be singled out by the integration $\oint U \sin\theta\, d\theta$. Again it is necessary to check that the part $v_2(r) \cos 2\theta$ of the velocity does not contribute to this integral. Instead of the boundary condition on $Z$ being $Z'(a) = 0$, $Z$ must now be matched to a solution $Z = Ar + C/r$ in the walls, with the usual boundary conditions (2.31). But $aZ'(a) = Z(a)$ and we find that $A = C = 0$, i.e. only the axisymmetric part of $v$ contributes to $\oint U \sin\theta\, d\theta$. Hence we can apply all the equations (2.33) to (2.36) to this case, where $S$ is now interpreted as the quantity $\oint U \sin\theta\, d\theta / bB\pi v_{\mathrm{m}}$, the integral being taken round the *outside* of the pipe.

This result opens up new possibilities for circular transverse-field meters of invariant sensitivity, provided one is prepared to approximate to the quantity $\oint U \sin\theta\, d\theta$ by using a sufficient number of peripheral electrode tappings. A small number would probably suffice for most practical purposes, particularly when the walls were thick.

## 2.3. Flowmeter 'end-shorting'

All the preceding §2.2 has dealt with the two-dimensional problem; it has been assumed that conditions vary at an insignificant rate along the pipe and in particular that the ends of the flowmeter and the edges of the magnetic field are reasonably remote. This is not a realistic assumption for all practical flowmeters, since to provide a magnetic field over a great length of pipe would be extravagant. We therefore conclude this chapter with a study of some of the consequences of having the edges of the magnetic field sufficiently close to the electrodes to influence their performance. Einhorn* led the way to the solution of this problem.

The edges of the field will actually be fringe regions where the field falls off gradually. Each fringe region will extend in the $z$-direction for a distance of the same order of magnitude as the magnet gap in the

* Einhorn, H. D. (1940). *Trans. Roy. Soc. S. Afr.* **28**, 143.

case of transverse-field meters. In axial-current meters the falling off of the field would depend on the detail construction of the energising current circuit. Figure 19 shows how the magnetic field and e.m.f. at a given radius would fall to zero at the ends of a meter in which the axial current flows in the fluid. With axial-current meters care must be taken to dispose external current leads axisymmetrically or to have them sufficiently remote to have no effect on the field in the meter.

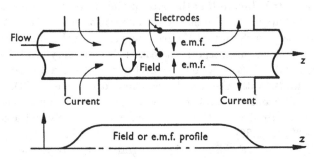

Fig. 19. Flowmeter with axial current in the fluid. The lower figure indicates how the circular magnetic field and radially induced e.m.f. fall off at the ends.

### 2.3.1. *Field with abrupt edge*

Figure 20 represents a view of a transverse-field meter, looking along the direction of the field. For simplicity we assume for a start that abrupt edges of the field occur at $z = \pm c$, the electrodes $X$ and $Y$ being at the middle of the flowmeter's length, where $z = 0$. Between its edges the field is assumed to be uniform, purely transverse and equal to $B$. Elsewhere it is zero. This neglect of fringing is most realistic for the case of a rectangular channel that is narrow in the field direction. Then the gap width and fringing distance can be small in comparison with the length $2c$ and depth $2b$ of the meter.

Figure 20 can also represent qualitatively the case of the axial-current meter if we imagine it rotated about the line $PQ$ to generate an annular flow channel or about the line $RS$ to generate a circular pipe. The representation is only approximate, because in axial-current meters the field varies with radius.

Hartmann* realised in 1937 that where the field falls off the induced e.m.f. in the fluid will also decrease, allowing short-circuit currents to flow at the edges of the field, somewhat as indicated in fig. 20. Unless the electrodes $X$ and $Y$ are remote from the edges

---

* Hartmann, J. (1937). *Math.-fys. Medd.* **15**, nos. 6 and 7.

of the field, a significant fall in $U_{XY}$ for a given flow rate will be produced by this process of 'end-shorting'. Conducting walls will assist the end-currents to flow, lowering $U_{XY}$ still further. In the case of an axial-current meter with the energising current in the fluid, the end-currents would be superposed on this. In other words the imposed current distribution would be distorted by the fluid motion.

Further assumptions permit us to solve the problem posed in fig. 20. Having already studied fairly exhaustively the effect of non-uniformities of velocity we now ignore these and take the fluid

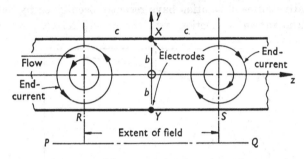

Fig. 20. Side view of flowmeter showing current loops at the edges of the field.

velocity $v$ as constant. If we also take the channel to be rectangular in cross-section, fig. 20 represents all planes of current and fluid flow and the problem is two-dimensional. In a circular pipe the problem is a three-dimensional one, but the general conclusions would be similar. We also assume initially that the walls are non-conducting so that the end-currents flow wholly in the fluid. Conducting walls are considered in §2.3.3.

Referred to the axes shown in fig. 20, the current flow and potential distribution obey the equations

$$j_z/\sigma = -\partial U/\partial z$$

and $$j_y/\sigma = -\partial U/\partial y \quad (+Bv \text{ if } |z| < c).$$

Because $\partial j_y/\partial y + \partial j_z/\partial z = 0$, it follows that $\nabla^2 U = 0$ everywhere, $v$ and $B$ being constant. Here $\nabla^2$ contains only $y$ and $z$ derivatives. By symmetry $U$ is constant when $y = 0$ and is taken as zero there. Then $U$ tends to zero when $|z|$ is very large also. At the walls when $y = \pm b$, $j_y$ must vanish, which specifies a boundary condition on $\partial U/\partial y$. $U$ and its derivatives are continuous across the edges of the field, $z = \pm c$.

37

Nevertheless the solution is given here in different forms for the regions inside and outside the field. For the range $|z| < c$,

$$U = Bvb \left[ \frac{y}{b} - \frac{8}{\pi^2} \Sigma \frac{(-1)^{(n-1)/2}}{n^2} \cosh \frac{n\pi z}{2b} \exp \left( - \frac{n\pi c}{2b} \right) \sin \frac{n\pi y}{2b} \right], \quad (2.43)$$

while for the range $|z| > c$,

$$U = Bvb \frac{8}{\pi^2} \Sigma \frac{(-1)^{(n-1)/2}}{n^2} \sinh \frac{n\pi c}{2b} \exp \left( - \frac{n\pi |z|}{2b} \right) \sin \frac{n\pi y}{2b}. \quad (2.44)$$

Each summation is taken over all odd positive integral values of $n$. Alternative forms of solution have recently been given by Sutton[*] and Fishman[†] in connection with the closely related problem of

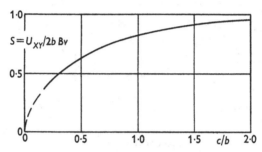

Fig. 21. Fall in sensitivity due to end-shorting (non-conducting channel, abrupt edge to field).

end-shorting in magnetohydrodynamic generators. That problem differs in that the walls $y = \pm b$ within the magnetic field are made conducting to permit the extraction of power from the fluid.

From (2.43) it emerges that

$$S = \frac{U_{XY}}{2bBv} = 1 - \frac{8}{\pi^2} \Sigma \frac{1}{n^2} \exp \left( - \frac{n\pi c}{2b} \right). \quad (2.45)$$

$S$ is plotted against $c/b$ in fig. 21, which shows clearly the loss of signal occasioned by end-shorting when the field edges are sufficiently near the electrodes. The analysis is meaningless as $c/b$ tends to zero owing to our neglect of fringing. Equation (2.45) indicates that $S$ reaches 0.99 when $c/b = 2.8$. For this and higher values of $c/b$ the effect of end-shorting may be regarded as unimportant when the walls are non-conducting. To eliminate end-shorting when $c/b$ is not large it may

  [*] Sutton, G. W. (1959). *General Electric Report* R 59 SD 431.
  [†] Fishman, F. (1959). *AVCO-Everett Research Note* 135. See also: Birzvalk, Y. A. (1959). *Latv. PSR Zinat. Akad. Vēstis.* No. 12, 49. (end-shorting in electromagnetic pumps). Boucher, R. A. & Ames, D. B. (1961). *J. Appl. Phys.* **32**, 755.

sometimes be worth while to insert streamwise insulating vanes or baffles in the fluid so as to bar the circulation of the currents at the edges of the field.

One incentive to select a small value of $c/b$ is that if it is desired to make the output signal as large as possible, but only a fixed amount of magnetic flux is available, $c$ should be made as small as possible. The raised value of $B$ more than compensates for the increased end-shorting. This follows from (2.45).

### 2.3.2. Fringed field

The analysis above may easily be generalised to allow partially for fringing. To keep the problem two-dimensional we ignore any components of magnetic field in the $y$- or $z$-directions or any variation of the transverse field $B$ in its own direction, while permitting $B$ to

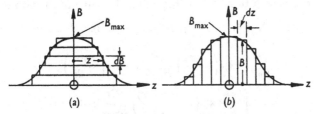

Fig. 22. Field profile with fringing.

vary with $z$ in the symmetrical manner shown in fig. 22. We continue to assume uniform flow through a rectangular channel with non-conducting walls. Note that the equation $\nabla^2 U = 0$ still applies.

This problem may be solved equally well by superposing the induced voltages due to elements of field $dB$ extending over an axial length $\pm z$, as suggested in fig. 22 $a$, or by superposing those due to elements of field $B$ extending over an axial length $dz$, as suggested in fig. 22$b$. Using the first approach and (2.43) gives

$$S = \frac{U_{XY}}{2bvB_{\max}} = 1 - \frac{8}{\pi^2} \Sigma \frac{1}{n^2} \int_0^{B_{\max}} \exp\left(-\frac{n\pi z}{2b}\right) \frac{dB}{B_{\max}}. \quad (2.46)$$

Using the second approach and (2.44) gives

$$S = \frac{4}{\pi} \Sigma \frac{1}{n} \int_0^\infty \frac{B}{B_{\max}} \exp\left(-\frac{n\pi z}{2b}\right) \frac{dz}{b}. \quad (2.47)$$

The results (2.46) and (2.47) are identical; knowing the field profile $B(z)$ we may evaluate $S$ from either.* We shall here be content to

* For alternative solution that assumes $B$, a periodic function of $z$, see: Wyatt, D. G. (1961). *Phys. Med. Biol.* 5, 289, 369.

derive some instructive general results concerning the desirability of fringing, which might be deliberately accentuated by the use of shaded poles or a variable air gap.

The question to settle is this: if a given amount of flux is available, is it desirable to have fringing or abrupt edges to the field? The answer is that having the *abrupt* edges gives the maximum voltage induced between the electrodes for given values of flow rate, total flux and peak flux density. Even though fringing tends to inhibit the end-currents, it entails some transfer of flux in a direction *away from* the electrodes. It is a general principle that the nearer the flux is to the electrodes the better. Another instance of this was given at the end of the preceding section.

To prove that abrupt edges are best for a given amount of flux, we need to show that (2.46) gives a smaller sensitivity than (2.45), where $c$ is chosen to make the total flux the same in the two cases. This requires that $\int_0^{B_{\max}} (z - c)\, dB/B_{\max} = 0$. It will suffice to show that

$$\int_0^{B_{\max}} \exp\left(-\frac{n\pi z}{2b}\right) \frac{dB}{B_{\max}} > \exp\left(-\frac{n\pi c}{2b}\right),$$

or that

$$\int_0^{B_{\max}} \exp\left(-\frac{n\pi(z-c)}{2b}\right) \frac{dB}{B_{\max}} > 1.$$

But this follows from the fact that

$$\exp\left(-\frac{n\pi(z-c)}{2b}\right) \geqslant 1 - \frac{n\pi}{2b}(z-c).$$

Another way of expressing this result is as follows. Supposing a sharp-edged uniform field is already available and it is proposed to add extra flux at the edges, should this be fringed (i.e. non-uniform) or not? The answer is no; the extra flux will contribute most strongly to the output signal if it is all added at the full strength of the existing field. Deliberate fringing of the field is a poor way of improving flowmeter performance.

To see the importance of fringing, consider a particular example. In a circular or square tube the fringing region will extend a distance of the order of $2b$ axially. To represent this case we may take a flux distribution of the form shown in fig. 23, in which $c$ is chosen so as to permit a comparison with an abrupt field case having the same total flux. Obviously $c$ must be larger than $b$ here. Applying (2.46) gives

$$S = 1 - \frac{16}{\pi^3} \Sigma \frac{1}{n^3} \sinh \frac{n\pi}{2} \exp\left(-\frac{n\pi c}{2b}\right).$$

40

This result is plotted in fig. 24, to which has been added a dotted curve from (2.45) representing the abrupt field case. It is clear that fringing intensifies the end-shorting for a given total flux. We have now to go to $c/b = 3\cdot04$ to secure $S = 0\cdot99$, in contrast to the value $c/b = 2\cdot8$ in the abrupt field case.

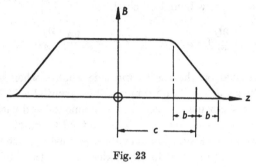

Fig. 23

It is not possible at the present stage of knowledge to say how accurately the above formula for $S$ could be applied to circular pipes. A reasonable course would be to use the mean semi-depth $\pi a/4$ in place of $b$ for a pipe of radius $a$.

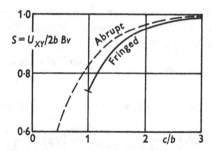

Fig. 24. Fall in sensitivity due to end-shorting showing the extra loss due to fringing (non-conducting channel).

### 2.3.3. *The effect of conducting walls*

Conducting walls will intensify the loss of signal due to end-shorting. This loss is additional to that due to short-circuiting via the side walls of the tube which occurs in the two-dimensional case in the absence of edges to the field. As this effect has already been discussed, we shall now ignore it and concentrate on the extra losses associated with axial currents in the top and bottom walls of a flowmeter at the edges of the field.

41

The first case to consider is a generalisation of the one studied in §2.3.1. Again an abrupt edge to the field is assumed, but now the walls $y = \pm b$ are taken to be thin, but conducting. Their effect is charac- terised by the dimensionless group $d = w\kappa/b\sigma$, $w$ and $\kappa$ being the wall thickness and conductivity respectively. The only modification to §2.3.1 needed is a new boundary condition

$$\frac{\partial U}{\partial y}(-Bv \quad \text{if} \quad |z| < c) = \pm \frac{w\kappa}{\sigma} \frac{\partial^2 U}{\partial z^2}. \qquad (2.48)$$

The positive sign on the right-hand side applies when $y = b$, the negative when $y = -b$. Condition (2.48) is essentially (2.39), modi- fied to allow for $v$ being non-zero at the boundary and with contact resistance omitted. Contact resistance will tend to inhibit end-short- ing via the walls, and leaving it out enables us to estimate end-shorting in the worst case, provided the electrodes contact the fluid directly. Now the generalised versions of (2.43) and (2.44) are

$$U = Bbv\Sigma A \exp(-k|z|/b)\sin ky/b, \quad \text{if} \quad |z| > c,$$

and $\qquad U = Bbv\{y/b - \Sigma C \cosh kz/b \sin ky/b\}, \quad \text{if} \quad |z| < c,$

where the summations extend over all the positive roots $k$ of the equation $kd \tan k = 1$. There are no complex roots. The coefficients $A$ and $C$ are found to be given by

$$C \exp kc/b = A \operatorname{cosech} kc/b = 2\left(\frac{\sin k/k^2 - \cos k/k}{1 - \sin 2k/2k}\right),$$

since the eigenfunctions $\sin ky$ are orthogonal. It follows that the sensitivity

$$S = 1 - 2\Sigma\left(\frac{(\sin k)^2/k^2 - \sin 2k/2k}{1 - \sin 2k/2k}\right)\exp\left(-\frac{kc}{b}\right). \qquad (2.49)$$

This formula fails as $c \to 0$. The result (2.49) is exhibited in fig. 25, which shows how a significant amount of wall conductivity soon lowers the sensitivity to intolerable levels. Though the assumptions that the walls are thin and the field edge is abrupt are not realised in practice, these results should be correct in order of magnitude.

A case of some interest is that where $d$ is large, but $c/b$ is also large so that $S$ is not disastrously low. This might occur in an axial-current meter of the annular type, where $2b$ now corresponds to the annular gap. To secure a good averaging effect, the wall conductivity would be high and $d$ large. We are assuming that the axial and circumferential

42

wall conductivities are the same, i.e. the walls are not laminated. In this case only the first term of the series in (2.49) is needed and then

$$S = 1 - \exp(-c/bd^{\frac{1}{2}}). \qquad (2.50)$$

This result can be derived more directly from the assumptions appropriate to this case. Moreover, it is then easy to allow for contact resistance and so include the case in which the influence of contact

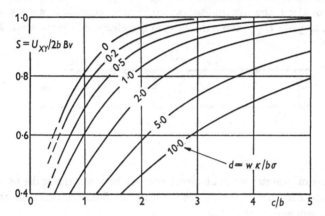

Fig. 25. The increase of end-shorting due to wall conduction.

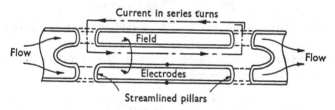

Fig. 26. Annular flowmeter with axial exciting current flowing in central conductors (one typical series turn shown).

resistance is most adverse, that of an annular axial-current meter having electrodes on the walls rather than in contact with the fluid and having direct contact between the inner and outer walls at the current supply points at the ends of the meter. Such contact is inevitable if the nature of the fluid dictates all-welded construction. This case appears in fig. 26.

If $d$ is large and the contact resistance $\tau$ is not so large as to inhibit end-shorting via the conducting walls severely, the axial part of the end-current loops will tend to flow almost entirely in the walls rather than the fluid. The obvious approximations to make are that in the fluid the current flows only in the $y$-direction as indicated in fig. 27 and that all variables depend only on the axial co-ordinate $z$. If $J$ and $U_w$ are the axial current per unit width and potential in the wall, respectively, then

$$j_y = dJ/dz, \quad J = -w\kappa dU_w/dz$$

and $U_w = Bbv$ (or zero outside field region) $-j_y(\tau + b/\sigma)$, if $U$ is taken as zero along the centre-line, $y = 0$. These equations are easily solved to yield $U_{XY}$, the p.d. between the electrodes in the wall. In

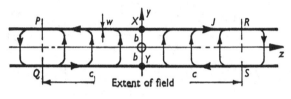

Fig. 27

the case of a meter without direct contact between the walls at its ends, the end-shorting takes place across the fluid lying outside the field region and we find that

$$S = \frac{U_{XY}}{2bBv} = 1 - \exp \frac{-c/b}{\{d(1 + \sigma\tau/b)\}^{\frac{1}{2}}},$$

which is a generalisation of (2.50). The conditions that $J$ and $U_w$ are continuous at the edges of the field have been satisfied.

If instead there are direct short circuits $P$–$Q$ and $R$–$S$ at the ends of the meter, the result becomes

$$S = 1 - \operatorname{sech} \frac{c/b}{\{d(1 + \sigma\tau/b)\}^{\frac{1}{2}}},$$

and the short-circuiting is twice as drastic as in the previous case at large values of $c/b$. For the loss of signal to be made less than 1 per cent we must have

$$\frac{c/b}{\{d(1 + \sigma\tau/b)\}^{\frac{1}{2}}} > 5 \cdot 3.$$

In the absence of contact resistance this implies

$$(c/bd^{\frac{1}{2}})^2 = c^2\sigma/wb\kappa > 28,$$

a condition easily satisfied with liquid metals. However, if contact resistance is severe, as may be the case with mercury and stainless steel for instance, $\sigma\tau/b$ may be much larger than unity (e.g. as large as

44

100), and then end-shorting is very marked for practicable values of $c/b$. Although contact resistance inhibits the circulation of the end-currents, it still lowers the voltage appearing between electrodes *in the walls*. In contrast, if the electrodes have direct access *to the fluid*, contact resistance reduces the loss of signal due to end-shorting.

For flowmeters in which $S$ does not reach the value unity even in the two-dimensional case in the absence of end-shorting, the values of $S$ given by the various formulae in this §2.3 should be used as multiplying factors to correct the 'two-dimensional' values of $S$ deduced in the earlier sections of this chapter. This, if not exact, is a reasonable approximate procedure. The real point is that any flowmeter for which $S$, as given in §2.3, departs drastically from unity must be calibrated empirically. In view of the simplifying assumptions made, the results given in this section serve more to indicate trends and so to guide flowmeter design than to give data for the precise calibration of flowmeters.

### 2.3.4. *Integrated-voltage flowmeters*

Section 2.2.6 showed how suitably combining measurements of the potential all round the periphery enabled one to eliminate the effect of non-uniformity and asymmetry of the velocity profile, provided the magnetic field was uniform. It might be expected that an extension of this technique could eliminate the effect of non-uniformity of the transverse field, i.e. the effect of end-shorting.

Consider the two-dimensional problem in the $(y, z)$-plane treated in the previous sections. Let there be flow at a uniform velocity $v$ in the $z$-direction in a rectangular channel with non-conducting walls under a field $B$ that is purely perpendicular to the $(y, z)$-plane. Fringing in both directions is approximated by taking $B$ as a function of $y$ and $z$.

We may integrate the equation

$$j_y/\sigma = -\partial U/\partial y + Bv$$

over that part of the $(y, z)$-plane that lies within the channel, noting that $\int_{-\infty}^{\infty} j_y \, dz = 0$ in this case, and taking $\sigma$ as constant.

As a result
$$\int_{-\infty}^{\infty} \Delta U dz = \phi v, \tag{2.51}$$

where $\Delta U$ is the p.d. between electrodes at the top and bottom of the channel at each value of $z$ and $\phi$ is the total magnetic flux that passes through the fluid. $\phi$ is presumably known or easy to find. The integral on the left-hand side does not have to be taken over an infinite stretch in practice because $\Delta U$ falls effectively to zero outside the magnet in the course of a few diameters. The integral could be evaluated experimentally with reasonable precision from readings taken at a small number of electrodes along the top and bottom of the channel. One way of combining the readings would be to use two passive networks of resistance high compared with the fluid but low compared with the voltage-measuring device (see fig. 28). Then (2.51) would yield the velocity from the voltage measurement.

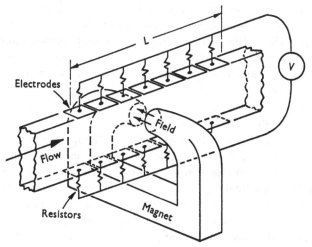

Fig. 28. Integrated-voltage flowmeter. $\displaystyle\int_{-\infty}^{\infty}\Delta U dz$ would be approximated by (voltmeter reading) $\times$ (length $L$).

This technique is certainly worth trying out as a means of creating a flowmeter with a predictable calibration in the face of drastic end-shorting, even when there is fringing in the $y$-direction also. Figure 28 illustrates an extreme case of this type. It would be necessary to investigate the effect of non-uniformity of velocity and curvature of the field lines. If the velocity were non-uniform, the device in fig. 28 would record merely the velocity between the poles.

When the pipe is circular and non-conducting and the velocity profile is axisymmetric, $v(r)$, it is possible to carry the analysis further provided the field $B$ can be assumed to be purely in the $x$-direction and now a function of $z$ only. Then the equation for the potential $U$ is

$$\nabla^2 U = \operatorname{div} \mathbf{v} \times \mathbf{B} = \mathbf{B} \cdot \operatorname{curl} \mathbf{v} - \mathbf{v} \cdot \operatorname{curl} \mathbf{B} = B(z)\, v'(r) \sin\theta,$$

if a dash denotes differentiation with respect to $r$. The solution is of the form $U = V(r, z) \sin \theta$, where

$$\frac{\partial^2 V}{\partial z^2} + \frac{\partial^2 V}{\partial r^2} + \frac{1}{r}\frac{\partial V}{\partial r} - \frac{V}{r^2} = Bv'. \qquad (2.52)$$

As $z \to \pm \infty$ outside the field, $U$, $V$ and $\partial V/\partial z$ tend to zero, and

$$\int_{-\infty}^{\infty} \frac{\partial^2 V}{\partial z^2}\, dz = 0.$$

If $\qquad Z(r) = \int_{-\infty}^{\infty} V dz \quad$ and $\quad C = \int_{-\infty}^{\infty} B dz,$

then (2.52) gives $\qquad Z'' + Z'/r - Z/r^2 = Cv'.$

Also $\partial V/\partial r$ and $Z'$ vanish at the wall. The analogy with (2.30) indicates that $Z(a) = Cav_{\mathrm{m}} = \frac{1}{2}\phi v_{\mathrm{m}}$, or that $\displaystyle\int_{-\infty}^{\infty} \Delta U dz = \phi v_{\mathrm{m}}$, in which $\Delta U$ is the p.d. between electrodes at the ends of the diameter of the pipe that is perpendicular to the field at each value of $z$. The integral may be evaluated experimentally as before and used to indicate $v_{\mathrm{m}}$.

This result provides a promising way of overcoming the problem of end-shorting in circular flowmeters provided the field does not deviate too far from the assumed form $B_x(z)$ i.e., is not too curved and is without fringing in the $y$-direction. The result may easily be generalised to include wall conductivity.

This concludes our survey of the theory of induction flowmeters in which the field and the velocity distributions may be regarded as known. The next chapter investigates the extent to which the field and flow patterns are deformed by the large electric currents that are induced in liquid metal flowmeters.

# Chapter 3

## EFFECTS PECULIAR TO LIQUID METALS

### 3.1. Introductory

The conductivities of liquid metals are several orders of magnitude higher than for even the best non-metallic liquid conductors (see Appendix for typical magnitudes). One immediate result of this is that the operation of a.c. flowmeters may be complicated by the occurrence of the skin effect, which would mean that an applied a.c. field, though uniform in the absence of the fluid, would be no longer uniform inside the fluid and might not penetrate to the interior of the fluid at all in extreme cases. Fortunately there is less incentive to use an a.c. field with liquid metals than with electrolytic conductors because there is no tendency towards polarisation at the electrodes and because the high thermal conductivity precludes large temperature differences that might have caused troublesome d.c. thermo-electric potentials. These could only be separated from the useful signal by the use of a.c. Nor do significant voltaic d.c. potentials normally occur. It is therefore common practice to use d.c. fields for liquid metal flowmeters, particularly since a permanent magnet may be used and a stabilised power supply for an electromagnet is then no longer required. The main advantage of a.c. operation which is forfeited is the ease of amplifying a.c. signals.

The motion of a liquid metal in an electromagnetic flowmeter raises magnetohydrodynamic problems, which involve the two major characteristic phenomena of magnetohydrodynamics:

(a) the induced currents distort the imposed magnetic field, and

(b) under the influence of the field, they produce body forces which affect the fluid dynamics, perhaps disrupting the velocity profile and adding to the pressure drop across the meter.

With non-metallic conducting fluids, these effects are negligible. In this chapter we shall assume that the imposed field $B$ can be treated as a d.c. one, but it should be noted that similar effects occur with a.c. fields. For instance, the dynamic effects tend to be proportional to $B^2$ and thus do not average out to zero when $B$ is alternating. This needs stressing as there are contrary statements in the literature.

This chapter studies the influence of the two phenomena (a) and

48

(b) on the performance of liquid metal flowmeters. The second one is more important and occupies most of the chapter. We shall dispose of the other phenomenon first.

## 3.2. Magnetic effects

The extent to which the imposed field is upset by the currents induced by the fluid motion is determined by Maxwell's equations for the steady state

$$\operatorname{curl} \mathbf{B} = \mu j \tag{3.1}$$

and

$$\operatorname{curl} \mathbf{E} = 0, \tag{3.2}$$

and by Ohm's Law

$$\mathbf{j} = (\mathbf{E} + \mathbf{v} \times \mathbf{B}). \tag{3.3}$$

The fluid permeability $\mu$ is assumed equal to its value *in vacuo*. From (3.1) to (3.3) it follows that

$$\operatorname{curl}(\mathbf{v} \times \mathbf{B}) + \lambda \nabla^2 \mathbf{B} = 0, \tag{3.4}$$

in which $\lambda = 1/\mu\sigma$, the diffusivity of magnetic fields through the fluid. This equation expresses the conflict between *convection* of magnetic field (the first term), which arises because of the well-known reluctance of the flux linked by a closed, highly conducting loop to change, and *diffusion* (the second term), for which Ohmic dissipation is responsible. The equation is closely analogous to the similar equation that occurs in the dynamics of viscous, non-conducting fluids and which expresses the balance between the convection and viscous diffusion of vorticity. In this case the extent to which convection prevails over diffusion, so producing boundary layers, etc., is measured by the familiar Reynolds number, a dimensionless group of the form $v$ (a typical velocity) times $l$ (a typical length) divided by $\nu$ (the kinematic viscosity, which is the diffusivity for vorticity or momentum). By an obvious analogy the quantity $vl/\lambda$ or $\mu\sigma vl$ has come to be called the magnetic Reynolds number. Its significance is that its magnitude measures the extent to which convection of field prevails over diffusion, i.e. the extent to which an imposed field is swept away by the fluid motion in the face of its tendency to revert, diffusively, to its unperturbed configuration. Thus $\mu\sigma vl$, which we denote by $R_m$, measures the severity of the perturbation of the imposed field by the induced currents.

The significance of $R_m$ may also be arrived at by a simpler approach. If we suppose the imposed field $B$ is not too drastically upset, the induced current densities will be of order of magnitude $\sigma Bv$ in view of (3.3). The three terms in (3.3) are usually of the same order of magnitude. If a current of density of this order circulates in a region having a length scale $l$, it follows from (3.1) (or the equivalent

$\oint \mathbf{H}.d\mathbf{r}$ = current linked) that induced fields of order $\mu\sigma Bvl$ will appear. The fractional perturbation of the imposed field is therefore again seen to be $\mu\sigma vl$ or $R_m$. This quantity can approach unity in large sodium flowmeters and then distortion of the imposed field may indeed be appreciable. In axial-current flowmeters with the current in the fluid, where there is an imposed *current* rather than an imposed field, $R_m$ measures the extent to which this imposed current pattern, together with its associated magnetic field, is swept downstream by the fluid.

In studying the manner in which the induced currents distort the imposed magnetic field of a flowmeter it is again convenient to adopt the stratagem of first considering, on a two-dimensional basis, a situation that is remote from the ends of the meter and invariant in the direction of flow, and then to correct this by considering the end effects which arise because of the finite extent of the magnetic field. In short meters the two effects will be completely intermingled and in no way distinguishable.

### 3.2.1. *Two-dimensional theory*

We consider first the transverse-field meter. It was seen in ch. 2 that the variation in velocity and induced e.m.f. over each cross-section produces circulating currents somewhat as shown at the top of fig. 29. These will produce an induced field $B_z$ parallel to the flow, distorting the imposed field lines in the manner indicated at the bottom of the figure, rather as if the fluid was dragging them down-stream like reeds in a river. There is no tendency to alter the imposed $x$-component of the field, because it does not link the current loops, nor to produce $z$-components of field outside the walls of the meter. We are assuming that no significant amount of current is drawn from the electrodes $X$ and $Y$ into an external circuit. Because the field outside the meter is not perturbed, no net force results on the magnet or coils that provide the imposed field. In the insulated wall case (fig. 29a) the electromagnetic forces act and react wholly within the fluid but if the walls conduct (fig. 29b) there is an electromagnetic force opposing the fluid motion which reacts on the walls, in addition to the viscous stress between walls and fluid.

The important question here is whether the presence of the induced field affects the flowmeter performance. Being parallel to the flow it plays no part in the induction term $\mathbf{v} \times \mathbf{B}$ and the flowmeter is unaffected. Equivalently it was remarked in §2.1 that (2.9) holds if the induced current and curl $\mathbf{B}$ are perpendicular to the motion.

50

Very similar conclusions apply to the case of axial-current meters at points remote from the ends where the problem is two-dimensional. The only difference is that in the likely event of an axisymmetric velocity profile there is no induced current flow and so no induced field.

It therefore appears that in neither type of flowmeter does the induced field affect the performance, if the ends of the meter are reasonably remote. What happens when this is not the case is discussed next.

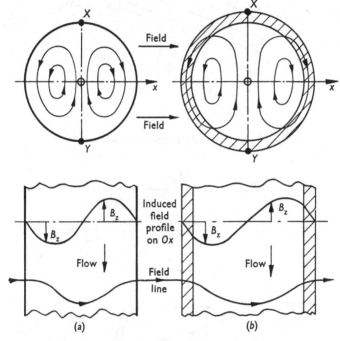

Fig. 29. Induced currents and fields in transverse-field meters: (a) non-conducting walls; (b) conducting walls.

### 3.2.2. *Edge effects*

In the last chapter it was seen that extra induced currents circulate in planes parallel, rather than perpendicular, to the direction of motion at the edges of the magnetic field of a flowmeter. This is illustrated in fig. 30. As Einhorn* remarked these currents will also produce an induced field, a distortion of the imposed field that prevails in the absence of motion. The severity of this effect will be measured

* Einhorn, H. D. (1940). *Trans. Roy. Soc. S. Afr.* **28**, 143.

by a magnetic Reynolds number, suitably defined, but now the problem is complicated by the fact that flux passing through the external magnet yoke links the induced currents.

If we take the simple situation illustrated in fig. 30, where the flowmeter has a rectangular cross-section that fills the magnet gap and assume that the reluctance of the yoke is negligible, we may readily

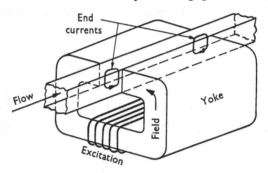

Fig. 30. End-currents in transverse-field meter.

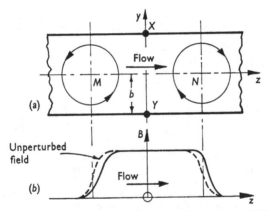

Fig. 31. (a) Side view of end-currents in transverse-field meter.
(b) Distortion of transverse field B.

estimate the distortion of the field. We assume that the applied current which energises the magnet is constant and ignore the actual three-dimensional or fringed nature of the edge field. Provided the field $B$ is not perturbed too greatly, the edge-current density is of order $\sigma B v$, circulating about the two fringe regions $M$ and $N$ (see fig. 31 a), at which the field perturbation will be a maximum. A field line passing round the yoke and through $M$ will link a demagnetising current of

52

order $\sigma Bvbg$, where $b$ is the semi-depth of the channel and $g$ is the gap. Thus the flux density at $M$ will be reduced by an amount of order $\mu\sigma Bvb$ or by a fraction of order $\mu\sigma vb$, the magnetic Reynolds number based on $b$.

The current distribution which follows from the analysis of end-shorting given in §2.3.1 may be used* to give the result that the fractional reduction of field at $M$ is $0\cdot37\mu\sigma vb$. This result is still approximate in view of the neglect of fringing. The reluctance of the magnet yoke would reduce the effect.

At the downstream edge the field will be augmented by a similar fraction. The whole phenomenon is reminiscent of armature reaction

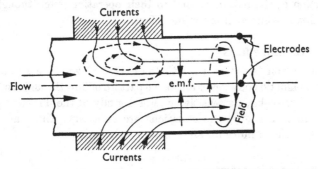

Fig. 32. End-currents in axial-current meter. The upper half shows the imposed currents and the induced currents (the dashed lines), the lower half shows the resultant currents.

in d.c. generators. The resultant field profile in the meter is illustrated in fig. 31$b$. Just as in §3.2.1, the fluid motion is tending to sweep the field downstream. The difference between the two situations is in the orientation of the current loops responsible for the distortion of the field.

Unless the meter is very short or $\mu\sigma vb$ unusually large, the field near the electrodes $X$ and $Y$ is not seriously perturbed and the flowmeter performance is not impaired. If $\mu\sigma vb$ became so large that the field near $X$ and $Y$ was severely deformed, the meter would become increasingly non-linear because the transverse-field distribution would vary with flow rate. Pfister, Dunham & Turner† have observed this effect.

Similar effects occur in axial-current meters. Figure 32 illustrates the upstream end of such a meter where the current is fed into the

* Shercliff, J. A. (1956). *J. Nucl. Energy*, **3**, 305.

† Pfister, C. G. & Dunham, R. J. (Oct. 1957). *Nucleonics*, **15**, 122. See also Turner, G. E. *U.S.A.E.C. Report* NAA-SR-4544 (April 1960).

liquid. The induced e.m.f. acts radially so as to oppose the spreading of the current from the wall into the fluid. The current lines tend to be swept downstream but the flowmeter performance will not be affected at normal values of the magnetic Reynolds number.

The perturbation of the magnetic field by the flow results in a net downstream force on the magnetic poles or the solid conductors that carry the energising current. This force itself may be exploited for the measurement of flow rate. Alternatively another type of flowmeter could operate by the direct measurement of the induced fields. These and other variants on the orthodox electromagnetic flowmeter are discussed in the next chapter. In them the magnetohydrodynamic distortion of the field is crucial to their operation, even though in orthodox flowmeters it is a minor effect.

### 3.3. Dynamic effects

This section is concerned with the extent to which the induced electromagnetic body forces affect the pattern of flow through a meter and, in particular, its sensitivity. For a study of the dynamic processes some further basic equations are necessary. These are the equation of continuity for an incompressible fluid

$$\text{div}\,\mathbf{v} = 0 \tag{3.5}$$

and the dynamic statement

$$\rho\frac{D\mathbf{v}}{Dt} + \text{grad}\,p = \eta\nabla^2\mathbf{v} + \mathbf{j}\times\mathbf{B}, \tag{3.6}$$

in which $\rho$, $p$ and $\eta$ are the fluid density, pressure and viscosity. Here the usual equation for incompressible viscous fluid flow gains the term $\mathbf{j}\times\mathbf{B}$, representing the electromagnetic force due to the induced currents in the prevailing magnetic field. Electrostatic forces on the charges existing in the fluid are always negligible in liquid metals. All the terms in (3.6) represent forces per unit volume. The acceleration term $D\mathbf{v}/Dt$ is an abbreviation for $\partial\mathbf{v}/\partial t + (\mathbf{v}.\text{grad})\,\mathbf{v}$, $D/Dt$ denoting time differentiation relative to an observer moving with the fluid.

The fluid-dynamical boundary condition will be that $\mathbf{v} = 0$ at stationary solid boundaries.

When liquid metal encounters the transverse field of a flowmeter and experiences the $\mathbf{j}\times\mathbf{B}$ forces, its behaviour in general may be split into three stages:

(*a*) at the upstream edge of the field the end-currents disturb the flow;

54

(b) then under the action of the uniform field in the middle of the meter the flow undergoes a settling towards

(c) an ultimate state, invariant in the direction of flow, which may or may not actually be reached within the finite length of transverse field available in the meter.

There will also be dynamic effects at the downstream edge of the field and a subsequent settling process outside the field.

We shall deal with the three stages in reverse order, first considering the ultimate state—a two-dimensional problem—and then the question as to when it would be reached in flowmeters and finally the dynamic edge effects.

### 3.3.1. *The ultimate state (two-dimensional theory)*

It is now assumed that all conditions have become independent of the downstream co-ordinate $z$, apart from a pressure gradient $\partial p/\partial z$ (and a potential gradient in axial-current meters with the current in the liquid). The analysis is restricted to laminar flow, but this is not so unrealistic as in ordinary pipe flows because we are postulating the presence of a magnetic field extensive and strong enough for the flow to have reached its ultimate state. Then the well-known tendency of a magnetic field to suppress turbulence in liquid metals* is likely to have taken effect. Some remarks on turbulent motions appear later in this chapter.

In any case it is useful to have the results appropriate to laminar flow because this is where the distortion of the velocity profile and the alteration of the flowmeter's sensitivity by the magnetic field is most marked. A turbulent flow generally gives a value of $S$ that lies between the unperturbed and the ultimate laminar values.

In the ultimate state the induced currents circulate in planes perpendicular to the direction of motion. The magnetic field, on the other hand, has components perpendicular to and parallel to the motion, and therefore there will be $\mathbf{j} \times \mathbf{B}$ forces also parallel and transverse to the motion.

The transverse $\mathbf{j} \times \mathbf{B}$ forces are produced only by the streamwise induced field $B_z$. They do not produce transverse motions because they can be balanced by transverse pressure gradients, being irrotational. This follows from the fact that $\mu\mathbf{j} = \operatorname{curl} \mathbf{B}_z$, so that

$$\mathbf{j} \times \mathbf{B}_z = (\mathbf{B}_z \cdot \operatorname{grad}) \mathbf{B}_z/\mu - \operatorname{grad}(B_z^2/2\mu),$$

* Hartmann, J. & Lazarus, F. (1937). *Math.-fys. Medd.* **15**, no. 7. Murgatroyd, W. (1953). *Phil. Mag.* **44**, 1348.

and $(\mathbf{B}_z.\text{grad})$ vanishes in the ultimate state because there is no variation in the $z$-direction. Since the fluid is incompressible and has no free surface these transverse pressure gradients are immaterial.

In circular axial-current meters with an energising current in the fluid there are transverse $\mathbf{j} \times \mathbf{B}$ forces due to this imposed current and its associated field. Again these result merely in transverse pressure gradients. This is simply the well-known pinch effect* attributable to the tendency of parallel, like currents to attract, and resulting in a maximum of pressure at the centre of the fluid. The pressure difference between the centre-line and the wall is equal to $\mu j^2 a^2/4$, in which $j$ is the axial-current density in the fluid and $a$ is the radius of the pipe. This is usually a very small pressure difference of the order of a centimetre head but might cause cavitation through self-constriction of the stream in extreme circumstances.

We turn next to the axial or streamwise $\mathbf{j} \times \mathbf{B}$ forces associated with the imposed field and the induced currents, taking first the simpler case of circular axial-current meters. With these the ultimate state is obviously axisymmetric and no induced currents flow, even when the walls are conducting. The ultimate state is merely the ordinary viscous velocity profile. The sensitivity is not affected by magnetic forces at all.

So far the $\mathbf{j} \times \mathbf{B}$ forces have exerted no influence on the sensitivity in the cases considered. The prospect is very different when we view the effect of streamwise $\mathbf{j} \times \mathbf{B}$ forces in transverse-field meters. In this discussion only a qualitative treatment will be given in most cases. For the reader who is interested in full and analytical details, references to the literature are noted.

Figure 33$a$ shows how the induced currents circulate in a transverse-field meter, producing magnetic forces which oppose the faster part of the motion and assist the slower part in accordance with Lenz's Law until this effect is limited through viscous action when the velocity gradient near the wall is increased. Figure 33$b$ indicates how the velocity profile changes. At the same time the increased viscous stress at the wall greatly increases the pressure losses for a given flow rate. If the walls are conducting the pressure loss is increased still further by a net electromagnetic force on the fluid, reacting on the walls.

When the field strength or conductivity is large enough the velocity is essentially uniform along $Ox$ in fig. 33 except for the boundary layers shown. Their thickness $\delta$ is easily estimated. The current density in the boundary layers is of order $\sigma B v_m$ since it is driven by a p.d. from top to

* Northrup, E. F. (1907). *Phys. Rev.* **24**, 474.

bottom of magnitude close to $Bv_{\mathrm{m}}(XY)$ and the opposing induced e.m.f. is small in the slow-moving boundary layer. Hence the magnetic force on the boundary layer per unit area of wall will be of order $\sigma B^2 v_{\mathrm{m}} \delta$, whereas the viscous drag per unit wall area is of order $\eta v_{\mathrm{m}}/\delta$, if we take $v_{\mathrm{m}}/\delta$ as a representative velocity gradient. These two forces must be comparable, i.e.

$$\sigma B^2 v_{\mathrm{m}} \delta \approx \eta v_{\mathrm{m}}/\delta,$$

and so
$$\delta = \frac{1}{B} \left(\frac{\eta}{\sigma}\right)^{\frac{1}{2}}. \tag{3.7}$$

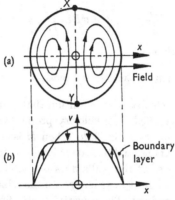

Fig. 33. (a) Induced currents in transverse-field meter.
(b) Resultant flattening of the velocity profile along $Ox$.

If $a$ is a typical dimension of the channel cross-section, the condition for the occurrence of boundary layers is clearly

$$a \gg \delta$$

or
$$M = Ba(\sigma/\eta)^{\frac{1}{2}} \gg 1,$$

which is usually satisfied in liquid metal flowmeters (see Appendix for typical magnitudes). The quantity $M$ is widely known as the Hartmann number after Hartmann* who was the first to solve a problem of this type, although Williams† was aware of its existence. It is evident that $M$ is the dimensionless quantity that determines the extent to which the ultimate profile is deformed by magnetic forces.

Hartmann took the case of motion between infinite parallel planes

* Hartmann, J. (1937). *Math.-fys. Medd.* **15**, no. 6.
† Williams, E. J. (1930). *Proc. Phys. Soc., Lond.*, **42**, 466.

in the presence of a uniform field $B$ spanning the gap. Then the currents and electric fields are purely in the $y$-direction, perpendicular to the directions of the velocity $(Oz)$ and of the uniform imposed field $(Ox)$, and all quantities depend only on the transverse co-ordinate $x$. From (3.2) it then follows that $E_y$ is constant. Ohm's Law (3.3) requires that

$$j_y = \sigma(E_y + vB)$$

while (3.6) yields

$$\frac{\partial p}{\partial z} = -Bj_y + \eta \frac{d^2v}{dx^2}.$$

Combining these equations leads to

$$\frac{d^2v}{dx^2} - \frac{\sigma B^2}{\eta} v = \left(\frac{\partial p}{\partial z} + \sigma BE_y\right) \Big/ \eta = \text{const} \qquad (3.8)$$

which has the solution

$$v = v_m\{\cosh M - \cosh(Mx/a)\}/\{\cosh M - \sinh M/M\}, \qquad (3.9)$$

$v_m$ being the mean velocity. The condition that $v$ vanish at the walls $x = \pm a$ has been satisfied. The velocity profile (3.9) for the case of $M = 10$ actually appears in fig. 33b. When $M$ is large, (3.9) predicts the appearance of exponential boundary layers at the walls. The conductivity or otherwise of the walls does not affect the form of the velocity profile although it does enter into the relation between flow rate and pressure gradient. The sensitivity of a flowmeter having a rectangular section, narrow in the field direction and therefore approximating to the Hartmann case, is given by (2.20), since the top and bottom walls are so narrow that it is immaterial whether they are perfectly conducting or not. Thus in this case the magnetic distortion of the velocity profile does not affect the sensitivity.

The pressure gradient is markedly affected by magnetic action, the formula being

$$-\frac{\partial p}{\partial z} = \sigma B^2 v_m \left\{\frac{d + \tanh M/M}{(1+d)(1 - \tanh M/M)}\right\}, \qquad (3.10)$$

which follows from (3.8), (3.9) and the fact that $-E_y = Bv_m/(1+d)$ (which is essentially (2.20)). The quantity $d$ or $w\kappa/a\sigma$ measures the importance of wall conductivity.

The ultimate state of flow is rather more complicated in circular or rectangular pipes under transverse fields. Boundary layers still occur when $M$ is large, but the more nearly parallel the wall is to the field, the greater is the boundary layer thickness.

58

In the case of non-conducting rectangular channels of width $2a$ in the field direction, the boundary layers on the walls that are parallel to the field and bear the electrodes prove to have thickness of order $a/M^{\frac{1}{2}}$. This is justified later. In contrast the side-wall boundary layer thickness is still of order $a/M$ as given by (3.7) for the Hartmann case and for the same reasons. A complete analysis of this case is available in the literature.* Outside the boundary layers the velocity is uniform and the corresponding value of $S$ is very close to unity, whatever the aspect ratio of the channel, if the walls are non-conducting and $M$ is large. This contrasts with the values up to and

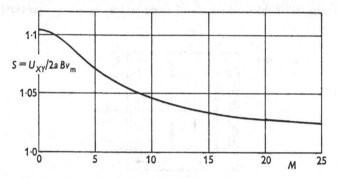

Fig. 34. Variation of $S$ with $M$ in square, non-conducting channel.

larger than (1·2) that occur with laminar flow in rectangular channels when $M$ is very small, as shown in fig. 16. Here there can be as much as a 20 per cent fall in sensitivity in the ultimate state as $M$ ranges from small to large values. Figure 34 shows this variation of $S$ with $M$ in the case of a square, non-conducting channel.†

The moral of this is that non-conducting rectangular channels of more or less square proportions should not be used if there is any danger of magnetic distortion or upstream disturbances causing these wide variations in sensitivity. Rectangular channels should only be used where they yield invariant values of $S$, either because they have highly conducting walls parallel to the field or because they are so narrow that the Hartmann case is approximated. Then the details of the velocity profile are immaterial, but the pressure losses are still

* Shercliff, J. A. (1952). *A.E.R.E.* (*Harwell*) *Report* X/R 1052; (1953). *Proc. Camb. Phil. Soc.* **49**, 136. See also: Birzvalk, Y. A. & Veze, A. (1959). *Latv. PSR Zināt. Akad. Vēstis*, 85. Chang, C. C. & Lundgren, T. S. (1959). *Heat Transfer and Fluid Mech. Inst.* 41. Ufland, Y. S. (1960), *Soviet Physics-Tech. Phys.* **5**, 1191 (transl.).

† Shercliff, J. A. (1952). *A.E.R.E.* (*Harwell*) *Report* X/R 1052.

of some interest. In the case of laminar flow in a narrow rectangular channel, the ultimate pressure gradient may be taken from (3.10). When $M$ is large this degenerates to

$$-\frac{\partial p}{\partial z} = \sigma B^2 v_{\mathrm{m}} \left\{ \frac{1}{M} + \frac{d}{1+d} \right\}. \tag{3.11}$$

Since the opposition to motion originates almost entirely by viscous shear in the side-wall Hartmann boundary layers (the first term on the right-hand side in (3.11)) and by the net magnetic force on the current flow between fluid and side walls, if conducting (the second term), (3.11) will still hold in channels of more or less square proportions

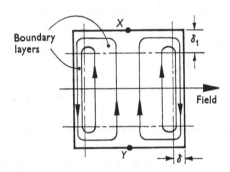

Fig. 35. Current circulation and boundary layers in rectangular flowmeter.

when $M$ is large. The quantity $d$ refers to the side-wall conductivity and thickness. Conducting walls parallel to the field merely assist the exchange of current between the core of uniform flow and the side boundary layers and walls.

It is easily shown why the top and bottom boundary layers have a thickness of order $a/M^{\frac{1}{2}}$ in rectangular, non-conducting channels. Figure 35 shows how the currents circulate, up in the core of uniform flow where the main e.m.f. of order $Bv_{\mathrm{m}}(XY)$ is induced, across the top and bottom layers of thickness $\delta_1$, say, and down the side boundary layers of thickness $\delta = a/M$. As the main resistance to current flow is in these thin side layers, the current density in them is still of order $\sigma Bv_{\mathrm{m}}$ and the total current in each layer per unit length of pipe is of order

$$\sigma Bv_{\mathrm{m}} a/M \quad \text{or} \quad v_{\mathrm{m}}(\sigma \eta)^{\frac{1}{2}}. \tag{3.12}$$

Thus $j_{\mathrm{y}}$ in the core and top and bottom layers is of order $v_{\mathrm{m}}(\sigma \eta)^{\frac{1}{2}}/a$ and the streamwise $\mathbf{j} \times \mathbf{B}$ force is of order $Bv_{\mathrm{m}}(\sigma \eta)^{\frac{1}{2}}/a$. This will be comparable with the pressure and viscous forces per unit volume in the top and bottom boundary layers, of order $\eta v_{\mathrm{m}}/\delta_1^2$, i.e.

$$Bv_{\mathrm{m}}(\sigma\eta)^{\frac{1}{2}}/a \approx \eta v_{\mathrm{m}}/\delta_1^2$$

and so
$$\delta_1 \approx a/M^{\frac{1}{2}}.$$

The vital point here is that the current distribution everywhere is largely determined by dynamic conditions in the side boundary layers.

The same principle applies to the case of circular channels* where boundary layers also occur as illustrated in fig. 36. The boundary layer thickness is inversely proportional to $M \cos\theta$, i.e. to the component of magnetic field normal to the wall. The balance between

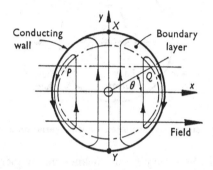

Fig. 36. Current circulation and boundary layers in circular flowmeter.

e.m.f.s and Ohmic drops and between viscous and $\mathbf{j} \times \mathbf{B}$ forces in the boundary layer also causes the velocity jump (from zero at the wall to $v_{\mathrm{c}}$ at the transition between the core and the boundary layer) to be proportional to the current flux down the boundary layer. The actual value of the total current in each layer per unit length of pipe is

$$v_{\mathrm{c}}(\sigma\eta)^{\frac{1}{2}}, \tag{3.13}$$

a resultant reminiscent of (3.12). The corresponding current in the wall in parallel with the boundary layer is $w\kappa/(\sigma a/M \cos\theta)$ or $dM \cos\theta$ times the value (3.13). $w$ and $\kappa$ are the thickness and conductivity of the wall while $a/M \cos\theta$ is the effective thickness of the boundary layer at a point where the normal is inclined at an angle $\theta$ to the field (see fig. 36). The quantity $dM \cos\theta$ is the ratio of the wall and boundary layer conductances.

* Shercliff, J. A. (1953). *Proc. Camb. Phil. Soc.* **49**, 136. (1956). *J. Fluid Mech.* **1**, 644. See also: Chang, C. C. & Lundgren, T. S. (1961). *Z. Angew. Math. Phys.* **12**, 100.

In the core the velocity $v_c$ becomes a function of $y$ only because any variation with $x$ is rapidly suppressed by eddy currents in the manner shown in fig. 37. In the core $j_y$ is uniform and equal to $(-\partial p/\partial z)/B$ so that the $\mathbf{j} \times \mathbf{B}$ and pressure forces can balance in the absence of significant viscous forces. Then $\partial j_x/\partial x = 0$ and so $j_x = 0$ by symmetry.

Figure 36 shows how the currents circulate up in the core and down the boundary layers and wall. Balancing current flows across the line $PQ$ gives

$$(PQ)\,(-\partial p/\partial z)/B = 2(\sigma\eta)^{\frac{1}{2}}v_c(1+dM\cos\theta). \qquad (3.14)$$

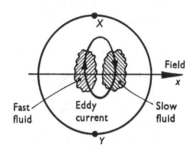

Fig. 37. Eddy currents suppressing velocity variation in $x$-direction.

If we neglect the boundary layer thickness in comparison with the length $PQ$, then $PQ = 2(a^2-y^2)^{\frac{1}{2}}$. With $\cos\theta = (a^2-y^2)^{\frac{1}{2}}/a$, (3.14) gives $v_c$ as the required function of $y$. This expression fails near $X$ and $Y$ but is valid elsewhere. Integrating $v_c$ over the cross-section, ignoring the small boundary layers, leads to

$$-\frac{\partial p}{\partial z} = \frac{3\pi B(\sigma\eta)^{\frac{1}{2}}v_m}{8a}\,(1+0{\cdot}883dM-0{\cdot}019(dM)^2...). \qquad (3.15)$$

This solution requires $M$ large but $dM$ not large in comparison with unity, i.e. $d$ must be small. The potential distribution and sensitivity $S$ may be found to the same accuracy by integrating the equation

$$\partial U/\partial y = Bv_c - j_y/\sigma \qquad (3.16)$$

along $XY$. The last term is negligible; the boundary layers and wall provide virtually all the resistance to current circulation. The resultant values for $S$ appear in fig. 38. $S$ takes the value $3\pi^2/32 = 0{\cdot}926$ as $d$ tends to zero (the case of insulating walls). This result shows that distortion of the velocity profile away from axisymmetry causes a fall in $S$ of 7 per cent, because flow is diverted from the regions of high $W$ to lower $W$ (see fig. 13). It must be remembered that fig. 38 is only valid for large values of $M$. As $M$ tends to zero the velocity profile becomes axisymmetric and $S$ tends to $1/(1+d)$, by (2.36) with $\tau = 0$.

Increasing the wall conductivity from zero increases $S$ at first when $M$ is large owing to the change in the velocity profile. Any decreasing tendency due to the short-circuiting effect of the conducting walls would stem from the term $j_y/\sigma$ which was omitted from (3.16).

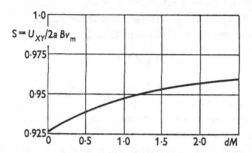

Fig. 38. Effect of wall conductivity on sensitivity at high $M(d = w\kappa/a\sigma)$.

When $dM$ becomes large the short-circuiting effect finally becomes dominant. Then the current-carrying capacity of the boundary layer is negligible and we may ignore its presence and concentrate entirely on the core. It is now easy to include the effect of uniform contact resistance $\tau$. It would also be easy to take the case of thick walls. The wall current per unit length of pipe (see fig. 12) is

$$J = -\frac{w\kappa}{a}\frac{dU_w}{d\theta} = -w\kappa\frac{dU_w}{dy}\cos\theta,$$

by Ohm's Law if the walls are thin. The suffices w and f distinguish potentials in the wall and the fluid. The currents in the core must permit $\mathbf{j} \times \mathbf{B}$ forces to balance the uniform pressure gradient and so again

$$j_y = (-\partial p/\partial z)/B, \quad \text{a constant.}$$

Equating the up and down currents gives $-J = aj_y \cos\theta$, while equating the radial current at the wall to the tangential rate of change of $J$ gives

$$\frac{U_f - U_w}{\tau} = \frac{1}{a}\frac{dJ}{d\theta} = j_y \sin\theta. \tag{3.17}$$

In the fluid $dU_f/dy = Bv - j_y/\sigma$. $U_f$ is a function of $y$ only because

63

$j_{\mathrm{x}}(=-\sigma\partial U_t/\partial x)$ is again negligible in the core. Differentiating (3.17) and combining it with the other equations yields the result

$$v = \frac{1}{\sigma B^2}\left(-\frac{\partial p}{\partial z}\right)\left(1+\frac{1}{d}+\frac{\sigma\tau}{a}\right) = \mathrm{const} \div v_{\mathrm{m}}, \qquad (3.18)$$

except in the viscous layers. This surprisingly simple result occurs because of the fortuitious cancelling of $\cos\theta$ from the equations. This does not occur in the case of non-circular channels. The viscous contribution to the pressure gradient/flow relation (3.18) is negligible just as it is in (3.11) when $M$ is large and $d$ is not small.

From the same equations it emerges that the sensitivity

$$S = 1/(1+d(1+\sigma\tau/a)). \qquad (3.19)$$

Alternatively this follows from (2.36) because the velocity profile is axisymmetric apart from the unimportant boundary layer regions. This value of $S$ corresponds to electrodes on the wall and not in direct contact with the fluid.

Braginskii* has discussed the case where $d$ is very large, but this would yield a worthless flowmeter, in view of (3.19).

For low values of $M$ it is possible† to solve the governing equations in terms of ascending power series in $M$. The results follow.

$$\text{Thin wall } (w \ll a) \quad S = \frac{1}{1+d(1+\sigma\tau/a)} - \frac{M^2}{576}\frac{1}{1+3d(1+3\sigma\tau/a)} + \dots, \qquad (3.20)$$

$$\text{Thick wall } S = \frac{2a^2}{(a^2+b^2)+(\kappa/\sigma)\,(1+\sigma\tau/a)\,(b^2-a^2)}$$

$$-\frac{M^2}{576}\frac{2a^4b^2}{(a^6+b^6)+(\kappa/\sigma)\,(1+3\sigma\tau/a)\,(b^6-a^6)}\dots.$$

In each case the first term coincides with the results (2.36) and (2.33) for general axisymmetric flow. At low values of $M$, magnetic distortion is seen to lower $S$ and having conducting walls reduces this effect. The next term in (3.20) is $+47M^4/921\,600$ if the walls are non-conducting.

These low-$M$ results are not important because at low values of $M$ the ultimate state would never be attained in a flowmeter of reasonable length.

* Braginskii, S. I. (1960). *Soviet Physics JETP*, **10**, 1005 (transl.). See also: Chang, C. C. & Lundgren, T. S. (1961). *Z. Angew. Math. Phys.* **12**, 100.
† Shercliff, J. A. (1952). *A.E.R.E.* (*Harwell*) Report X/R 1052; (1953). *Proc. Camb. Phil. Soc.* **49**, 136.

Enough results for the ultimate state of flow in circular pipes under transverse fields have now been given for the major trends to be apparent. The fact that (3.19) holds both at very low and very high values of $M$ might lead one to suppose that the sensitivity of circular flowmeters with conducting walls is independent of magnetic distortion of the velocity profile. This is certainly not true. Equation (3.20) shows how $S$ starts to descend below the value given by (3.19) as $M$ rises from zero, while if $d$ is small (so that $M$ may be large while $dM$ is not large) fig. 38 indicates how widely $S$ can vary with $M$.

### 3.3.2. Entry length

This section investigates the question as to whether the ultimate state would be reached within flowmeters of limited length. The point at issue is whether magnetic distortion of the velocity profile or upstream perturbations are likely to affect the sensitivity seriously. It is desirable that the ultimate state be reached in so far as that removes the influence of upstream disturbances, particularly in axial-current meters without highly conducting walls where only an axisymmetric state leads to a reliable value for the sensitivity; but undesirable in that for transverse-field meters the ultimate state yields deviant values for the sensitivity. In both types of meters magnetic forces hasten the settling of the flow to the ultimate state. Eddy currents occur in such a way as to eliminate inappropriate features of the velocity profile.

Figure 37 illustrated an example of this. Swirl, eddies or turbulence in the flow will also induce eddy currents and will be partially or completely damped during the settling process. We shall use the term *entry length* to describe the distance the fluid traverses after entry into the region of imposed magnetic field during the settling to the ultimate state.

In this section only laminar flow is considered. The transient behaviour of an initially turbulent flow is equally interesting but regrettably much less tractable. A linearized theory for the laminar entry process is available in the literature* but is too lengthy for inclusion in this book. Only the conclusions will be reproduced here, together with a qualitative discussion of the mechanics of the process.

Underlying the linearized theory is the restriction that the entry length be not too short, i.e. that quantities vary relatively slowly in the streaming direction. Then conditions on each different cross-

---

* Shercliff, J. A. (1956). *Proc. Camb. Phil. Soc.* **52**, 573; (1956). *J. Fluid Mech.* **1**, 644.

section of the pipe may be analysed on a two-dimensional basis, with the velocity and induced field still predominantly axial. At each cross-section the potential distribution, current flow, induced field etc., are determined by the two-dimensional quasi-steady theory given in §2.2. Then a flowmeter in which the ultimate state has not been reached develops the output signal that is appropriate to the velocity profile prevailing at the electrodes, with suitable allowance for end-shorting, if necessary.

At low values of $M$ the settling process still relies largely on viscous forces, and entry lengths are of the same order of magnitude as for non-conducting fluids. Consider the decay of the discrepancy between an arbitrary initial velocity profile and the ultimate state. The inertia forces per unit volume associated with a discrepancy $v'$ (in order of magnitude), decaying in a settling time $T$, will be of order $\rho v'/T$ per unit volume, which must be comparable with the associated excess viscous forces of order $\eta v'/l^2$, where $l$ is a dimension characteristic of the velocity discrepancy. Thus we see that

$$T \approx \rho l^2/\eta.$$

Multiplying by $v_m$, the mean velocity of flow, indicates that the corresponding entry length is

$$L \approx \rho l^2 v_m/\eta = R l^2/a,$$

where $R$ is the Reynolds number based on the radius $a$. We are only interested in the most persistent parts of the discrepancy in velocity. These will clearly be the ones with the largest length scale $l$. Figure 39 shows the discrepancy between an initially uniform velocity profile and the ultimate parabolic profile for a non-conducting fluid. The most persistent Fourier component of this

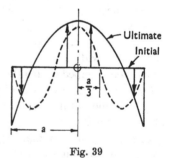

Fig. 39

discrepancy (shown dotted) has a length scale $l$ of order $a/3$, indicating an entry length of $Ra/9$. This estimate is reasonably compatible with the results of more exact analyses* for this case, where in a length $Ra/20$ or thereabouts the discrepancy decays in the ratio $1:e$.

In flowmeters $R$ is likely to be in the range $10^4$ to $10^5$. Clearly the entry length at low values of $M$ will be so large that virtually no

---

* Goldstein, S. (ed.) (1938). *Modern Developments in Fluid Dynamics*, vol. 1, p. 299 et seq. Oxford University Press.

significant change of the velocity profile will take place before the electrodes are reached. We therefore now turn to the case of high $M$, using arguments similar to those employed in the simple introductory case just considered. We shall assume that wall conductivity, if present, is not so large as to affect the order of magnitude of the entry length. The condition for this is that $dM$ be not very large in comparison with unity.

*Transverse-field meters.* The simplest case is Hartmann flow between parallel planes under a transverse field, corresponding to a narrow rectangular flowmeter. Consider the settling of a velocity distribution which is initially dependent only on the transverse co-ordinate $x$. Where the velocity is initially uniform the entry process simply entails the growth of the Hartmann boundary layers, the core velocity being virtually unchanged, if $M$ is large. Thus the boundary layers are isolated from one another and the wall spacing is irrelevant. In the growing boundary layers the excess current density is of order $\sigma v'B$, where $v'$ is a representative velocity discrepancy. The excess magnetic force $\sigma v'B^2$ will be comparable with the inertia force $\rho v'/T$, so that

$$T \approx \rho/\sigma B^2 \quad \text{and} \quad L \approx \rho v_{\mathrm{m}}/\sigma B^2 = Ra/M^2,$$

which is independent of wall spacing, as expected, and also viscosity. This result is confirmed by the more exact analysis.* The same arguments and result apply to the decay of other initial velocity discrepancies (dependent on $x$ only) including broad variations in the core of the flow where viscosity would be expected to play no role. Steeper variations of velocity which were affected by viscosity would decay more rapidly, but it is the most persistent discrepancies which concern us here. With typical values of $R$ around $10^5$ and $M$ around 100 it is evident that magnetic distortion could well occur in a narrow rectangular flowmeter with a length from inlet to electrodes equal to a few times the width $2a$. However, in such a meter the sensitivity $S$ is independent of velocity profile.

The entry length for flow in rectangular channels of more or less square proportions or circular pipes (or indeed for Hartmann flow between parallel planes when there are initial velocity variations in the $y$-direction) turns out to be $M$ times larger at the approximate value $Ra/M$. The first stage of the settling process is the establishment of Hartmann boundary layers in the time $\rho/\sigma B^2$. Then the more leisurely adjustment of the core profile (and growth of the top and

---

* Shercliff, J. A. (1956). *Proc. Camb. Phil. Soc.* **52**, 573.

bottom boundary layers in the rectangular case) ensues. In the discussion of the ultimate state it was seen that the magnetic force in the core was of order $Bv_c(\sigma\eta)^{\frac{1}{2}}/a$. This arose because the current distribution was controlled by the side boundary layers. Similarly in the settling process the excess magnetic force associated with the most persistent part of a decaying core velocity discrepancy of order $v'$ is of typical magnitude $Bv'(\sigma\eta)^{\frac{1}{2}}/a$, and this must be comparable with $\rho v'/T$. Thus $T \approx \rho a/B(\sigma\eta)^{\frac{1}{2}}$ and $L \approx \rho a v_m/B(\sigma\eta)^{\frac{1}{2}} = Ra/M$, as stated above. This result is confirmed by the more exact analysis.[*] The entry length for a circular pipe tends to be rather shorter than for a square pipe. Experimental evidence is that at electrodes at a distance $\frac{1}{2}Ra/M$ past the entrance to the region of transverse field the sensitivity $S$ is within 1 per cent of its ultimate value which corresponds to the fully distorted velocity profile.

If $R = 10^5$ and $M = 10^2$ to $10^3$, the entry length is evidently too large for significant magnetic distortion or magnetic damping of initial disturbances of the velocity profile to take place in a flowmeter of practicable length. But at low flow rates the magnetic forces would have time to act, with consequent variations in the sensitivity of the meter. If this were likely to occur in a given application it would be wise to replace the offending meter by one of the designs whose sensitivity is independent of velocity profile.

The presence of conducting walls would hasten the settling process by promoting the eddy currents that damp out the velocity discrepancies. The effect will be marked when the wall conductance is much larger than that of the Hartmann boundary layers, i.e. when $dM \gg 1$. Then the excess current density associated with a velocity discrepancy $v'$ would be of order $w\kappa v'B/a$ in the core, which leads to a value

$$L = \rho a v_m/w\kappa B^2,$$

for the entry length, which is $dM$ times smaller than for non-conducting walls. Thus significant magnetic distortion is to be expected in flowmeters with sufficiently highly conducting walls.

*Axial-current meters of circular cross-section.* The main difference in this case is that, though magnetic forces may participate in the settling process, the ultimate state is undistorted by magnetic forces. Moreover, any axisymmetric velocity profile gives the same value for the sensitivity. So the only important question concerns the entry length required for asymmetries of the velocity profile due to upstream

[*] Shercliff, J. A. (1956). *Proc. Camb. Phil. Soc.* **52**, 573; (1956). *J. Fluid Mech.* **1**, 644.

disturbances to decay. These asymmetries would affect the sensitivity unless the meter had highly conducting walls that could exert an averaging effect.

The settling of an axisymmetric velocity profile would only be affected by magnetic forces if turbulence was present.

The settling of an asymmetric profile will inevitably proceed rather differently from the transverse-field cases already discussed because there are no walls perpendicular to the field and so no Hartmann boundary layers able to control the current distribution any longer. Figure 40 illustrates part of one of the decaying modes or components of the velocity profile. The discrepancy between an initial velocity

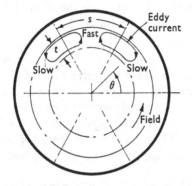

Fig. 40. Cross-section of axial-current meter.

profile and the ultimate one may be split into modes, each character-ised by its own exponential decay rate. The cells in the figure indicate regions where the fluid is moving faster or slower than it will ulti-mately, generating eddy currents more or less as shown. The crucial problem is what size and shape of cell gives the slowest decay. Making the cell larger in all directions makes decay due to viscosity slower. Making the cell longer in the $\theta$-direction or narrower in the radial direction inhibits the eddy currents and makes decay due to magnetic forces slower. This occurs because making the cell narrower radially reduces the excess e.m.f. (induced radially) while making it longer in the $\theta$-direction makes the eddy currents travel further. Thus the slowest decay occurs with cells of maximum $\theta$ extent and radial extent chosen as a compromise to make the viscous and magnetic damping rates equally weak. The maximum $\theta$ extent cannot be $2\pi$, because that corresponds to axisymmetric profiles, and is in fact $\pi$, since fast and slow regions alternate. The length $s$ will therefore be of the same order as $a$.

For cells of dimensions $s$ and $t$, as shown in fig. 40, the main resistance to current circulation is in the $\theta$-direction, and is of order $s/\sigma t$ per unit length of pipe. The excess e.m.f. is of order $Bv't$, where $v'$ is a representative velocity discrepancy, the total current is of order $(Bv't) \div s/\sigma t = \sigma Bv't^2/s$ and the current density is of order $\sigma Bv't^2/s^2$ when flowing radially. Thus the $\mathbf{j} \times \mathbf{B}$ force is of order $\sigma B^2v't^2/s^2$, and, for this to be comparable with the viscous forces so as to give the slowest decay, we need
$$\sigma B^2 v' t^2/s^2 \approx \eta v'/t^2.$$

If we put $s \approx a$ this gives $t \approx a/M^{\frac{1}{2}}$. Setting the $\mathbf{j} \times \mathbf{B}$ or the viscous forces approximately equal to the inertial force $\rho v'/T$ gives $T \approx a\rho/B(\sigma\eta)^{\frac{1}{2}}$. Then the entry length
$$L \approx Ra/M,$$

just as for transverse-field meters, but for very different reasons. $M$ should be based on a representative value of $B$ such as that prevailing at half-radius (or half way across the gap in annular meters).

The conclusion is that only at low flow rates will asymmetries of an inlet velocity profile decay significantly in axial-current meters. Therefore care should be taken to eliminate upstream disturbances unless the meter has highly conducting walls and is insensitive to velocity profile.

The presence of highly conducting walls would incidentally strongly promote the decay of asymmetries of the velocity profile by the damping action of eddy currents.

To summarise briefly: it appears that in most typical flowmeter applications with liquid metals, very little change of velocity profile will occur within the meter and the output signal will be at the mercy of upstream effects, unless the sensitivity is independent of velocity profile. However, at low flow rates it is more likely that upstream effects will be suppressed and the ultimate state reached. In transverse-field meters the sensitivity $S$ will vary according to the extent to which the ultimate state is reached.

### 3.3.3. *Edge effects*

Upsteam effects which might affect the performance of venturi meters and other orthodox flow-measuring devices may be avoided by having an adequate straight length of pipe ahead of the meter. In electromagnetic flowmeters* there is an 'upstream' effect which

* Shercliff, J. A. (1956). *J. Nucl. Energy*, **3**, 305.

is inherent in the device itself and so cannot be avoided. This is the perturbation of the velocity profile by the end-currents which circulate at the upstream edge of the magnetic field (see figs. 31 *a*, 32). In axial-current meters these only produce axisymmetric perturbations which do not affect the sensitivity, but in transverse-field meters the effect can be serious. In addition, in both types of meter the end-currents produce pressure losses which may indeed constitute the main pressure drop across the meter. Similar effects occur in electromagnetic pumps* and magnetohydrodynamic generators.†

In order to analyse this problem for transverse-field meters we make the assumptions that the flow channel is rectangular with nonconducting walls and that viscous effects are negligible. Then the problem is a two-dimensional one in the $(y, z)$-plane. The justification for neglecting viscous terms is that the dynamic process under discussion takes place over a short length of channel, whereas the viscous settling processes considered in the previous section usually occupy a great length of channel. The edge effect may be regarded as rapidly setting up a deformed velocity profile which then decays towards the ultimate state during the subsequent entry length. Even turbulence will not interfere with the edge effect very much because the forces involved can be stronger than the turbulent Reynolds stresses.

We also assume that the magnetic field $B$ is purely in the transverse $x$-direction and that it is not seriously distorted by the flow, i.e. $R_m$ is low. $B$ may be a function of the axial co-ordinate $z$, to allow partially for fringing.

If viscous or turbulent shear forces are neglected, in steady flow

$$\rho(\mathbf{v}.\mathrm{grad})\,\mathbf{v} = \mathbf{j} \times \mathbf{B} - \mathrm{grad}\,p. \qquad (3.21)$$

Eliminating $p$ by taking the curl of (3.21) gives, as the $x$-component,

$$\rho(\mathbf{v}.\mathrm{grad})\,\omega_x = -j_z\,dB/dz \qquad (3.22)$$

because div $\mathbf{v}$ and div $\mathbf{j} = 0$, with $\mathbf{v}$ and $\mathbf{j}$ flowing in $(y, z)$-planes and $\mathbf{B}$ in the $x$-direction. $\omega_x$ is the vorticity component $\partial v_z/\partial y - \partial v_y/\partial z$. Equation (3.22) shows how vorticity varies along a streamline. It is created only where $B$ is varying in the presence of current flow in the $z$-direction. The reason for this is that the consequent variation in the $y$-component of the $\mathbf{j} \times \mathbf{B}$ force tends to spin the fluid. If the velocity ahead of the flowmeter is uniform, vorticity appears as the fluid crosses the fringe of the field and is then conserved along each streamline as

---

* Rossow, V. J. (1960). *Rev. Mod. Phys.* **32**, 987.

† Sutton, G. W. & Carlson, A. W. (1961). *J. Fluid Mech.* **11**, 121.

the fluid rapidly settles to a pseudo-ultimate state of rectilinear flow, before the viscous forces have time to operate. This first settling process is obviously not affected by the $\mathbf{j} \times \mathbf{B}$ forces once $\mathbf{B}$ is uniform (by (3.22)) although the pressure distribution will be strongly affected wherever $\mathbf{j}$ occurs.

*Abrupt edge.* The simplest case is that where the fringing region is so narrow in comparison with the depth $2b$ of the channel that the field may be taken as having an abrupt edge. Then the vorticity $\omega_x$ changes

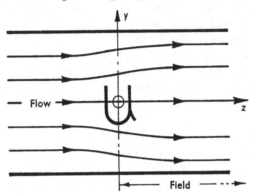

Fig. 41. Streamlines at upstream edge of field.

very rapidly with $z$. It can only change at a finite rate with $y$, however. Thus (3.22) approximates to $\rho v_z \, d\omega_x/dz = -j_z \, dB/dz$ in the edge region, and in the limit of an abrupt edge we get

$$\rho v_z \omega_x = -j_z B, \tag{3.23}$$

because $v_z$ and $j_z$ are continuous across the edge. Here $\omega_x$ is the vorticity after the edge has been crossed. Ahead it is assumed zero. Across the edge of the field the quantities $p$, $v_y$, $v_z$ and $\partial v_z/\partial y$ are concontinuous but $\omega_x$ and $\partial v_y/\partial z$ change abruptly. The motion takes the form shown in fig. 41, where the gradients of the streamlines are continuous but their curvature changes at the edge. The pressure is high in the region $U$ and the fluid approaching it is retarded. The motion leaving $U$ is further retarded by the $\mathbf{j} \times \mathbf{B}$ forces. The streamlines soon become parallel again but the vorticity created at the edge persists and results in a non-uniform velocity. Note that the velocity is raised in the regions near the electrodes which, in a circular transverse-field meter, tends to raise the sensitivity $S$ in view of §2.2.4.

It is easy to find a first-order, linearized solution of the problem by assuming that the fractional perturbations of the velocities are small. We let $v_z = v_m$ (the undisturbed, uniform velocity)$+ w$, and treat $v_y$ and $w$ as small quantities, neglecting quadratic terms in them. This is equivalent to neglecting the sideways displacement of the streamlines and implies that $\omega_x (= \partial w / \partial y - \partial v_y / \partial z)$ is conserved along the lines $y = \text{const}$ beyond the edge of the field. The function $\omega_x(y)$ is determined by the value of $j_z$ at the edge. This being a first-order theory we can take $j_z$ from the calculation in §2.3.1, where the velocity distribution was assumed unperturbed. Equation (2.44) enables us to find $j_z$ as $-\sigma \partial U / \partial z$ when $z = -c$. If the two edges of the field are reasonably remote from one another they do not interact and we may let $c \to \infty$, with the result that

$$-j_z = \frac{2\sigma v_m B}{\pi} \sum \frac{(-1)^{(n-1)/2}}{n} \sin \frac{n\pi y}{2b}, \quad (n \text{ odd}).$$

Hence
$$\omega_x(y) = \frac{2\sigma B^2}{\pi \rho} \sum \frac{(-1)^{(n-1)/2}}{n} \sin \frac{n\pi y}{2b}, \tag{3.24}$$

to the first order, by (3.23).

The problem is best solved in terms of a perturbation stream function $\psi$, where $v_y = -\partial \psi / \partial z$ and $w = \partial \psi / \partial y$, so that the equation of continuity is satisfied. Then

$$\nabla^2 \psi = 0 \quad (\text{if } z < 0) \quad \text{or} \quad \omega_x(y) \quad (\text{if } z > 0), \tag{3.25}$$

if we choose an origin of axes as in fig. 41. Equations (3.25) must be solved subject to the conditions $\psi = 0$ as $z \to -\infty$, $\psi$ and $\partial \psi / \partial z$ continuous at $z = 0$, and $\psi = 0$ at $y = 0$ and $\pm b$. As $z \to +\infty$, (3.25) becomes
$$d^2 \psi / dy^2 = \omega_x(y)$$

and so
$$\psi = \frac{8\sigma B^2 b^2}{\pi^3 \rho} \sum \frac{1}{n^3} \left( \frac{y}{b} - (-1)^{(n-1)/2} \sin \frac{n\pi y}{2b} \right), \tag{3.26}$$

where the constants of integration have been fixed by the boundary conditions. Hence as, $z \to \infty$,

$$w = \frac{\sigma B^2 b}{\rho} \left\{ \frac{8}{\pi^3} \sum \frac{1}{n^3} - \frac{4}{\pi^2} \sum \frac{(-1)^{(n-1)/2}}{n^2} \cos \frac{n\pi y}{2b} \right\}, \tag{3.27}$$

which is the expression for the perturbation of the velocity profile after it has become rectilinear again but before the viscous settling process has proceeded significantly. The curve labelled 'abrupt edge' in fig. 42 shows how the velocity profile is perturbed, the fluid being retarded near the centre of the channel and accelerated at the top and bottom near the electrodes. The velocity change at the centre

73

is of order $\sigma b B^2/10\rho$ and the fractional change is $\sigma B^2 b/10\rho v_{\mathrm{m}}$. $\sigma B^2 b/\rho v_{\mathrm{m}}$ is another of the dimensionless groups that are important in magneto-hydrodynamics (see Appendix for typical magnitudes). It measures the relative magnitude of the $\mathbf{j} \times \mathbf{B}$ and inertial forces, i.e. the ability of the magnetic field to modify an inviscid flow. Thus the velocity perturbation can be very serious in practice, particularly with liquid sodium for which $\sigma$ is high and $\rho$ low. As soon as the fractional perturbation becomes larger than, say, 20 per cent this first-order analysis becomes unrealistic.

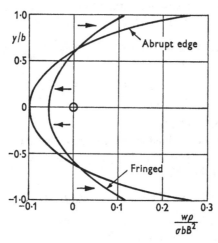

Fig. 42. Perturbation of velocity profile by edge effects.

A similar additive perturbation occurs at the downstream edge of the field, doubling rather than eliminating the perturbation.

It is important to know roughly how long the flow takes to become rectilinear again. This emerges from the solution of (3.25), which may easily be found if the expression (3.26) is Fourier-analysed into terms of the form $A_{\mathrm{m}} \sin \dfrac{m\pi y}{b}$ ($m$ even and odd). Then to each of these components there correspond terms of the form

$$\tfrac{1}{2} A_{\mathrm{m}} \exp \frac{m\pi z}{b} \sin \frac{m\pi y}{b} \quad (z < 0),$$

or
$$A_{\mathrm{m}} \left\{ 1 - \tfrac{1}{2} \exp\left( -\frac{m\pi z}{b} \right) \right\} \sin \frac{m\pi y}{b} \quad (z > 0),$$

in the solution for $\psi$. The value $m = 1$ gives the term which most concerns us here, the one that takes longest to reach the steady rectilinear state as $z$ increases. It is apparent that this has virtually been reached when $z$ equals a low multiple of $b/\pi$, $2b$ say. Thus the rectilinear state is attained very rapidly, at least for the weak interaction considered here. There is evidence from the closely related experiments of Rossow* that when the fractional velocity perturbations are large the rectilinear motion gives place to eddying motions which would be most undesirable in a flowmeter. No systematic experiments on dynamic edge effects in flowmeters have yet been performed, but they are urgently called for, particularly since it is far from obvious what exactly happens in circular pipes as distinct from the rectangular channels considered here. Probably some swirl is generated.

*Fringed edge.* The more realistic problem of the fringed edge with $B$ a function of $z$ may also be solved to the same degree of approximation. To the first order (3.22) gives

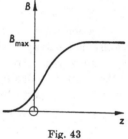

Fig. 43

$$\rho v_{\mathrm{m}} \frac{d\omega_{\mathrm{x}}}{dz} = -j_z \frac{dB}{dz}$$

and $\rho v_{\mathrm{m}} \omega_{\mathrm{x}}(y) = \int -j_z dB$, evaluated at constant $y$. Here $\omega_{\mathrm{x}}$ is the vorticity in the rectilinear state after the edge. The $j_z$ distribution due to the fringed field edge $B(z)$ shown in fig. 43 is given by

$$-j_z = \sigma \frac{\partial U}{\partial z} = \frac{2\sigma v_{\mathrm{m}}}{\pi} \Sigma \frac{(-1)^{(n-1)/2}}{n} \sin \frac{n\pi y}{2b}$$

$$\times \int_0^{B_{\max}} \exp\left(-\frac{n\pi|z-z'|}{2b}\right) dB(z'),$$

if we add all the contributions of jumps $dB$ located a distance $|z-z'|$ away from the point of interest, $z'$ being the variable of integration. Hence

$$\omega_{\mathrm{x}}(y) = \frac{2\sigma B_{\max}^2}{\pi\rho} \Sigma \frac{(-1)^{(n-1)/2}}{n} \sin \frac{n\pi y}{2b}$$

$$\times \int_0^{B_{\max}} \frac{dB(z)}{B_{\max}} \int_0^{B_{\max}} \exp\left(-\frac{n\pi|z-z'|}{2b}\right) \frac{dB(z')}{B_{\max}}, \quad (3.28)$$

$z$ being the variable in the extra integration. This should be compared with the result (3.24) for the abrupt edge. Because

$$\int_0^{B_{\max}} \exp\left(-\frac{n\pi|z-z'|}{2b}\right) \frac{dB(z')}{B_{\max}} < \int_0^{B_{\max}} \frac{dB}{B_{\max}} = 1,$$

* Rossow, V. J. (1960). *Rev. Mod. Phys.* **32**, 987.

75

the coefficients of the Fourier series (3.28) are individually less than in (3.24), which suggests that the velocity perturbations are *lessened* by fringing. This does not follow rigorously because the terms of the series alternate in sign. However, it is easily shown that the perturbation $w$ of the velocity profile at the walls $y = \pm b$ ultimately reaches the value

$$w = \frac{8\sigma B^2 b}{\pi^3 \rho} \Sigma \frac{1}{n^3} \int_0^{B_{\max}} \frac{dB(z)}{B_{\max}} \int_0^{B_{\max}} \exp\left(-\frac{n\pi|z-z'|}{2b}\right) \frac{dB(z')}{B_{\max}}, \quad (3.29)$$

which is evidently less than the corresponding value for the abrupt edge, namely,

$$w = \frac{8\sigma B^2 b}{\pi^3 \rho} \Sigma \frac{1}{n^3}.$$

That fringing reduces perturbation of the velocity profile by end-currents is confirmed if we take the plausible example of a field fringed as in fig. 23, representing the case of a flowmeter of square or circular cross-section where the magnet gap, fringe width and pipe diameter are all equal to $2b$ approximately. We assume that $c$ is large enough for the two edge regions not to interact and for the velocity perturbation at the upstream edge to be complete before the second edge is encountered. This perturbation is then

$$w = \frac{\sigma B^2 b}{\rho} \Sigma \left\{ \frac{16}{\pi^4 n^4} - \frac{8}{\pi^3} \frac{(-1)^{(n-1)/2}}{n^3} \cos\frac{n\pi y}{2b} \right\} \left\{ 1 - \frac{1-\exp(-n\pi)}{n\pi} \right\}, \quad (3.30)$$

which is plotted in fig. 42 for comparison with the abrupt edge case. The perturbations are seen to be reduced by a rough factor of 2 by fringing of this severity.

It is interesting that fringing is desirable from this aspect whereas it was undesirable in relation to end-shorting.

Fringing would not materially affect the length of pipe necessary for the velocity to reach its perturbed state.

Conducting walls, by tending to divert from the fluid to the walls the axial current $j_z$ at the edge of the field, will obviously reduce the perturbation of the velocity profile. Since an upper estimate of the effect is already available from the case of non-conducting walls, we shall not investigate this case, although to do so is not difficult.

### 3.3.4. *Pressure losses in flowmeters*

For efficient operation of a flow-circuit it is essential to know and minimize the pressure losses, to which an electromagnetic flowmeter may contribute. Only in the case of a liquid metal does the pressure drop in a flowmeter differ from the ordinary hydraulic losses associated with the flowmeter duct. These may be readily estimated from existing hydraulic data and need no discussion here.

When liquid metals pass through transverse-field flowmeters, three distinct types of excess pressure loss may occur, each one corresponding to one of the three preceding sections. We shall take them in the same order, which is the order of increasing importance.

*Pressure gradient in the ultimate state (laminar flow).* The relevant formulae which give the increased pressure gradient associated with the full distortion of the viscous velocity profile by $\mathbf{j} \times \mathbf{B}$ forces have already appeared in §3.3.1 (equations (3.10), (3.11), (3.15) and (3.18)). In the only situation where the ultimate state is likely to be reached— that where $M$ is large—all these formulae give a value for $(-\partial p/\partial z)$ of the order of $v_{\mathrm{m}} B(\sigma\eta)^{\frac{1}{2}}/a$ whatever the shape of the channel provided $dM$, representing the effect of wall conductivity, is not much larger than unity. Thus the pressure gradient is increased above the value of order $\eta v_{\mathrm{m}}/a^2$ (for non-conducting fluids) by the large factor $M$. If $dM$ is large and $d$ of order unity, the pressure gradient is of order $\sigma B^2 v_{\mathrm{m}}$, i.e. larger again by another factor $M$, unless contact resistance is significant.

*Pressure loss during the settling process.* Consider a circular pipe with non-conducting walls into which the fluid enters at uniform velocity. There is an extra pressure loss $\Delta p$ over the entry length in excess of the pressure loss over the same length of pipe for flow in the ultimate state. This occurs for two reasons:

(*a*) The creation of axial momentum. The changes of velocity at various parts of the pipe during the entry process are of the same order as, but rather less than, $v_{\mathrm{m}}$. The ultimate non-uniform velocity profile has higher momentum than the entry profile. Thus the associated contribution to $\Delta p$ is of order, but somewhat less than, $\rho v_{\mathrm{m}}^2$. However, in some situations the initial profile can have higher momentum than the final one and then the momentum contribution to $\Delta p$ is negative.

(*b*) The fact that the skin friction is higher than its ultimate value in the growing Hartmann boundary layers. The mean value of the excess friction will be comparable with but less than the ultimate value. Thus the associated contribution to $\Delta p$ is of order $(-\partial p/\partial z)_{\mathrm{ult}}$ times the entry length (of order $\rho v_{\mathrm{m}} a/B(\sigma\eta)^{\frac{1}{2}}$) and hence

$$\Delta p \approx \text{ but } < \rho v_{\mathrm{m}}^2$$

due to this effect also.

The result of a more elaborate but still approximate calculation*

* Shercliff, J. A. (1956). *J. Fluid Mech.* **1**, 644.

77

for this case of a circular pipe in a transverse field is that the total value of $\Delta p = 0.041\,\rho v_{\mathrm{m}}^2$. The evidence* is that the values of $\Delta p$ given by this method of calculation are underestimated by a factor of roughly 2, presumably because the method suppresses the contribution $(a)$ to $\Delta p$. A reasonable working rule would seem to be to take $\Delta p = \rho v_{\mathrm{m}}^2/10$ for the loss of pressure during a completed, laminar entry process in circular non-conducting pipes under transverse fields. Though the field $B$ does not appear in the result it exerts a strong effect. In the absence of $B$, $\Delta p$ is of order $0.7\rho v_{\mathrm{m}}^2$,† for settling from an initially uniform velocity profile. This result is of major interest here because of the fact that, after the fluid leaves a flowmeter in which any significant amount of profile distortion by magnetic forces has occurred (either by edge effects or by settling towards the ultimate state), a further settling process must take place in the downstream region outside the magnetic field. Because $\Delta p$ is larger in the absence of the field, the major contribution to the pressure loss due to settling processes apparently occurs after the meter. Since the velocity profiles far upstream and downstream of the meter are the same, the momentum contributions to the total $\Delta p$ cancel out. In view of this and failing better information, we will arbitrarily adopt the value $\tfrac{1}{2}\rho v_{\mathrm{m}}^2$ for the total $\Delta p$ due to inlet and outlet laminar settling processes whenever there is a significant amount of magnetic distortion of the flow and consequent settling.

For flow between parallel planes or in deep rectangular channels under transverse fields when the inlet velocity is uniform, the velocity only changes significantly in the thin boundary layers that occupy a fraction $1/M$ of the channel width. The entry length is less than it is for circular pipes by a factor $1/M$. For both these reasons $\Delta p \approx \rho v_{\mathrm{m}}^2/M$ in this case for a complete settling process, as more elaborate analysis* shows. However, here we are more interested in the settling of velocity profiles that have been perturbed by edge effects so that the velocity varies also in the $y$-direction, and then the settling lengths and pressure losses are much the same as for the circular case already discussed.

No information is available about likely values of $\Delta p$ in turbulent flows but one's guess is that it will not be very different. Obviously the momentum contribution to $\Delta p$ will still be of order, but less than $\rho v_{\mathrm{m}}^2$.

---

* Shercliff, J. A. (1956). *Proc. Camb. Phil. Soc.* **52**, 573.

† Goldstein, S. (ed.) (1938). *Modern Developments in Fluid Dynamics*, vol. 1, p. 299 et seq. Oxford University Press.

Some simple criterion as to how far the two settling processes and their associated $\Delta p$ occur in each case is necessary. The outlet settling process will always occur to an extent determined by the severity of the profile distortions at the inlet and outlet edges of the field and the degree of completeness of the inlet settling process within the limited length of the flowmeter.

The severity of the edge distortions is measured by the dimensionless quantity $\sigma B^2 b/\rho v_m$, which must exceed unity for significant distortion to occur. The degree to which the magnetically distorted ultimate state is approached if the flow is laminar is determined by the magnitude of the entry length $Ra/M$ in comparison with $2c$, the length of the transverse-field region. Distortion of this kind will be significant if the quantity $(Mc/Ra)$ or $\{B(\sigma\eta)^{\frac{1}{2}}/\rho v_m\}(c/a)$ approaches unity. Unless $c$ is very much larger than $b$, this quantity is smaller by the large factor $bM/c$ than $\sigma B^2 b/\rho v_m$, which therefore is the important parameter.

*End-current pressure loss (non-conducting walls).* Here again $\Delta p$ arises from two causes, the total change of momentum associated with a change in velocity profile and the $\mathbf{j} \times \mathbf{B}$ forces opposing the motion. $\Delta p$ is the pressure drop existing between two cross-sections, upstream and downstream of the upstream edge of the field, situated where the flow is rectilinear and $\partial p/\partial y = 0$. In the notation used in §3.3.3 we have

$$\rho v_m \frac{dw}{dz} = -j_y B - \frac{\partial p}{\partial z} \tag{3.31}$$

from (3.21), to the first order. Integrating (3.31) gives

$$\Delta p = \frac{1}{2b} \int_{-b}^{b} dy \int_{-\infty}^{\infty} j_y B(z)\, dz + \frac{\rho v_m}{2b} \int_{-b}^{b} w\, dy,$$

and the last term, representing the momentum contribution to $\Delta p$ (to the first order), vanishes by continuity. The total momentum contribution is negligible until $w$ begins to approach $v_m$. Equation (3.31) also implies that $\rho v_m w = \Delta p$ when $y = \pm b$ because $j_y = 0$ there. Here $w$ is the value attained in the rectilinear flow beyond the edge and is given by (3.29). $\Delta p$ is therefore reduced by fringing just as $w$ is. Fringing is often deliberately encouraged in d.c. pumps* and magnetohydrodynamic generators† in order to reduce the pressure losses due to the end-currents. The same principle applies to flowmeters.

* Blake, L. R. (1959). *J. Nucl. Energy*, pt. B (*React. Tech.*), **1**, 65. See also: Birzvalk, Y. A. & Tutin, I. A. (1956). *Trud. Inst. Fiz. Akad. Nauk, Latv. S.S.R.*, no. 8, p. 59. Watt, D. A. (1956). *Engineering, Lond.*, **181**, 264.

† Sutton, G. W. (1959). *General Electric Report* TIS R 59 SD 431.

The case where the edge of the field is abrupt provides the upper limit to $\Delta p$, equal to
$$\rho v_{\mathrm{m}} w(b) = 0.27\sigma B^2 b v_{\mathrm{m}}. \tag{3.32}$$

In the case of the linearly fringed field illustrated in fig. 23, representing roughly the case of a circular tube,
$$\Delta p = \rho v_{\mathrm{m}} w(b) = 0.11\sigma B^2 b v_{\mathrm{m}},$$

which is markedly less than for the abrupt edge.

A similar pressure loss occurs at the downstream edge of the field because of the $\mathbf{j} \times \mathbf{B}$ forces.

*End-current pressure loss (conducting walls).* Consider merely the extreme case where the wall conductivity is so high and $d$ or $w\kappa/\sigma b$ so large that the axial part of the end-currents flows almost entirely in the walls at the top and bottom of the channel. Little perturbation of the velocity profile occurs but the pressure losses are large. From an extension of the discussion of this case in §2.3.3 it follows that, for each edge,
$$\Delta p = B \int_0^\infty j_{\mathrm{y}} dz = \tfrac{1}{2}\sigma B^2 v_{\mathrm{m}} b \, d^{\frac{1}{2}},$$

if the field has abrupt edges that are sufficiently far apart for the two end zones not to interact. It has been assumed that there is no contact resistance. If there is direct contact between the two walls at the ends of the meter, as may occur in central-conductor, annular meters, the value of $\Delta p$ is doubled. Making $d$ large is seen to increase $\Delta p$ above the value given by (3.32) in a manner proportional to $d^{\frac{1}{2}}$.

*Axial-current meters.* Here the ultimate state and the approach to it in axisymmetric flow involves no extra pressure gradients due to magnetohydrodynamic effects. Moreover, the circular field usually changes rather slowly with axial distance at the ends of the meter and so the pressure losses due to the end-currents will be weaker than in transverse-field meters, particularly since field strengths are likely to be lower in axial-current meters. Finally, the more complicated geometric form of central-conductor meters of this type means that ordinary hydraulic losses are likely to be more important than magnetohydrodynamic ones.

*Pressure losses in transverse-field meters—a summary.* It is now possible to compare the three different kinds of pressure loss. The following table gives their orders of magnitude, the walls being assumed non-conducting. The tabulated values refer to the case of a square or circular pipe where fringing of the field has been allowed for approximately as in §3.3.3. $2c$ is the length of the region of transverse field in the direction of flow. The column entitled 'entry and exit loss' refers to the loss associated with the settling of the velocity profile and is only relevant when the quantity $\sigma B^2 b / \rho v_m$ exceeds unity. This table shows that the condition for the second column to be relevant is also the condition that this source of loss be not much greater than the end-current loss. It may never even be as large as the end-current loss. There is an obvious need for experiments to establish the relative magnitudes of the end-current loss and the exit settling loss more precisely than these rough estimates here permit.

| | End-current losses (both edges) | Entry and exit loss | Ultimate state loss (laminar flow) |
|---|---|---|---|
| | $0.22\sigma B^2 a v_m$ | $\tfrac{1}{2}\rho v_m^2$ | $2cv_m B(\sigma\eta)^{\frac{1}{2}}/a$ |
| In approximate ratio | 1 | $2\rho v_m/\sigma B^2 a$ | $(9/M)\,(c/a)$ |

The table also shows that, when $M$ is large, the ultimate state loss could only become significant at impracticably large values of $c/a$.

In the case of a deep rectangular channel ($b \gg a$) the only modification necessary in the first line of the table is to put $0.54\,\sigma B^2 b v_m$ for the end-current losses if the edges of the field are taken as abrupt.

Typical magnitudes of quantities appearing in the tables are exhibited in the Appendix.

### 3.3.5. *Turbulence and instability*

These are the least fully understood aspects of the fluid dynamics of liquid metals in flowmeters. The available theory is inadequate and empirical information is scanty. We can claim to know reasonably well what will happen in a flowmeter under given conditions, how the sensitivity will vary, etc., *but only if the flow is laminar*. In most practical situations the flow is turbulent upstream of the meter. Uncertainty surrounds the extent to which similar effects will occur with an initially turbulent flow as for a laminar one.

Some of the effects already discussed are essentially unaffected by turbulence provided one interprets them in terms of mean velocities. Edge effects associated with the end-currents at the edges of the field come into this category. Turbulence does not enter significantly into the problem any more than does viscosity since turbulence only acts on the mean flow like an extra, variable viscosity. Thus the distortion of the field by the flow, the pressure loss and the deformation of the velocity profile at the edges of the field are not much changed if the flow is turbulent. However, this profile deformation may have an effect in determining whether or not a flow becomes or stays turbulent, particularly if the deformation is strong enough to lead to eddies or back flow.

The entry length problem is unexplored in cases where turbulence is present initially or ultimately or both. One just cannot say how long the entry length is in these cases and whether the ultimate state would be reached inside a flowmeter in given circumstances. In so far as turbulence acts like an augmented viscosity it may tend to reduce entry lengths. On the other hand, if the entry process involves the partial or complete suppression of initial turbulence in the flow or alternatively the growth of instabilities which lead to turbulence in an initially laminar flow, the length necessary to reach the ultimate state may well be larger.

If the complete entry process requires a longer length of pipe than is available in the flowmeter (as will probably be the case in practice, except at low flow rates) the sensitivity of the flowmeter will depend on the upstream profile, modified perhaps by edge effects. In circular meters the upstream profile will be axisymmetric, whether the flow is laminar or turbulent, in the absence of close upstream disturbances. Then the appropriate values of $S$ will apply, if edge effects are unimportant.

If the field is strong enough and the flow slow enough, the ultimate state may be reached inside the flowmeter. These same conditions imply that laminar flow, or at least strongly modified turbulent flow, is likely to occur. Rather more information is available about the nature of the ultimate state than about other aspects of the turbulent flow problem, but there are still many obscurities. It has long been known[*] that a transverse magnetic field inhibits turbulence in channel flows and may completely eliminate it. At the same time the magnetic field renders a laminar flow more stable, as Lundquist remarked.[†] Lock[‡] has analysed this in detail, taking the case of Hartmann flow between parallel plane walls, a distance $2a$ apart, and making approximations appropriate to the case of a liquid metal. He emerges with the result (for high values of $M$) that

$$R_{\mathrm{crit}} = 50,000\,M, \tag{3.33}$$

where $R_{\mathrm{crit}}$ is the value of $\rho v_m a/\eta$ above which laminar flow becomes

[*] Hartmann, J. & Lazarus, F. (1937). *Math.-fys. Medd.* **15**, no. 7.

[†] Lundquist, S. (1952). *Ark. Fys.* **5**, 297.

[‡] Lock, R. C. (1955). *Proc. Roy. Soc.* A, **233**, 105. See also: Pavlov, K. B. & Tarasov, Y. A. (1960). *Appl. Math. Mech. Leningr.* **24**, 1079 (transl.).

unstable to infinitesimal disturbances. From our experience of ordinary hydrodynamics we expect the value of $R_{\text{crit}}$ observed in experiments to be much lower because finite disturbances are always present. $R_{\text{crit}}$ is bound to be a function of $M$ only in any situation where the magnetic Reynolds number is low enough for induced fields to be negligible. The reason that $R_{\text{crit}}$ is here *proportional* to $M$ is that when $M$ is large, the flow consists of two exponential Hartmann boundary layers isolated from one another by a core of uniform flow whose width is unimportant. Thus the channel width must cancel out from (3.33), as indeed it does. Lock found that the magnetic forces only served to determine the undisturbed flow and played no part in the mechanics of instability.

The theory of instability of flows in channels of other shapes has not been explored. Since the core velocity profile in high-$M$ flow in circular pipes is not unlike the parabolic profile for ordinary hydrodynamic flow between parallel planes, it might be expected that $R_{\text{crit}}$ be similar in these two cases, i.e. independent of $M$ or magnetic field. However, in the author's experience,[*] $R_{\text{crit}}$ for this case certainly rises with $M$, roughly according to the law

$$R_{\text{crit}} = 250\, M.$$

It may be that here magnetic forces do participate in the mechanics of instability. Wooler[†] points out that magnetic forces do not affect the stability of laminar flow between parallel planes if the field is transverse to the flow but parallel to the planes. This suggests that the stability of laminar flow in axial-current meters will not be affected by magnetic forces since the field is parallel to the walls in them also.

With flowmeters another problem is more important in practice: what is the ultimate state of an initially turbulent flow? This has been studied experimentally,[‡] in the case of flow between parallel planes, but in the more interesting case of flow in circular pipes not much is known. It is very probable that under a transverse field a turbulent profile is closer to being axisymmetric than is the distorted laminar profile, the reason being that turbulence raises the effective viscosity and lowers the effective value of $M$. But even this is not certain because of the markedly anisotropic effect of a magnetic field on turbulence[§] whereby it tends to damp only vorticity perpendicular to the

[*] Shercliff, J. A. (1956). *J. Fluid Mech.* 1, 644.

[†] Wooler, P. T. (1961). *Phys. of Fluids*, 4, 24.

[‡] Hartmann, J. & Lazarus, F. (1937). *Math.-fys. Medd.* 15, no. 7. Murgatroyd, W. (1953). *Phil. Mag.* 44, 1348.

[§] Lehnert, B. (1955). *Quart. Appl. Math.* 12, 321.

field direction. Reported results for liquid metal flowmeters in turbulent flows occurring in practical applications indicate that observed values of $S$ are often 3 or 4 per cent lower than the values appropriate to axisymmetric velocity profiles. In contrast the high-$M$, laminar value of $S$ is lower by 7 per cent. This tends to confirm the view that the ultimate turbulent profile in a magnetic field always leads to a value of $S$ intermediate between the values for the axisymmetric profiles and the high-$M$ laminar one, the exact value depending on the degree to which the turbulence is suppressed. The laminar results are then chiefly of interest as lower limits to $S$.

Before a discussion of the experimental results available for turbulent flows, some more general remarks are necessary. In liquid metals it is a good approximation to neglect induced fields in comparison with the uniform imposed field $\mathbf{B_0}$, particularly the *turbulent* induced fields (as distinct from those due to the mean induced currents). Then the only relevant dimensionless groups are $R = \rho v_m a/\eta$ and $M = Ba(\sigma/\eta)^{\frac{1}{2}}$ or combinations of them. Neglecting induced fields reduces the equations of magnetohydrodynamics to the simpler forms

$$\rho \left(\frac{\partial \mathbf{v}}{\partial t} + (\mathbf{v}.\mathrm{grad})\,\mathbf{v}\right) + \mathrm{grad}\,p = \mathbf{j} \times \mathbf{B_0} + \eta \nabla^2 \mathbf{v} \quad (\rho\ \text{const}),$$

and $$\mathbf{j} = \sigma(\mathbf{E} + \mathbf{v} \times \mathbf{B_0}),$$

plus the linear equations of electromagnetism. The only non-linear term is the inertial term $\rho(\mathbf{v}.\mathrm{grad})\,\mathbf{v}$ that occurs in ordinary hydrodynamics. Thus, if we take the mean values of all these equations, they retain their form unchanged in terms of the mean values of the quantities $\mathbf{v}$, $p$, etc., except for the usual Reynolds stress term which arises from the non-linear inertial term. The magnetic field affects the motion in two ways. One occurs even in homogeneous turbulence and is the direct, anisotropic damping of eddies by eddy currents. This reduces the Reynolds stresses and thus affects the mean motion if the mean velocity is not uniform. The other, which we might term the Hartmann effect, is the direct distortion of the mean velocity profile by the associated mean induced currents.

If the imposed field is parallel to the mean flow there is obviously no Hartmann effect and only the direct damping of eddies occurs. This case has attracted considerable attention* but is not relevant to flowmeters.

* Stuart, J. T. (1954). *Proc. Roy. Soc.* A, **221**, 189. See also: Bader, M. & Carlson, W. C. A. (1958). *N.A.C.A.* TN 4274. Drazin, P. G. (1960). *J. Fluid Mech.* **8**, 130. Globe, S. (1959). *Heat Transfer and Fluid Mech. Inst.*, 68. Globe, S. (1961). *Trans. Amer. Soc. Mech. Engrs*, **83**, series C, 445. Michael, D. H. (1953). *Proc. Camb. Phil. Soc.* **49**, 166. Tarasov, Y. A. (1960). *Soviet Physics JETP*, **10**, 1209, (transl.). Velikhov, E. P. (1959). *Soviet Physics JETP*, **9**, 848 (transl.).

Hartmann's[*] experiments revealed both effects. Applying a modest transverse field to a weakly turbulent flow resulted in a fall in pressure gradient for a given flow rate; the eddies were being inhibited and the Reynolds stresses reduced. Applying a strong enough field to affect a strongly turbulent flow resulted in a rise in pressure gradient; the wall shear stresses were being increased by the Hartmann effect.

Crausse & Poirier[†] have more recently done experiments on turbulent flow in circular pipes under transverse fields at low values of $M$.

The best available source of experimental data on flows between parallel planes is the work of Murgatroyd,[‡] discussed recently by Harris.[§] Murgatroyd found that initially turbulent flows became apparently laminar within the length of his apparatus (30–40 times the gap $2a$ between the planes) when $R/M < 225$, $M$ being large. Note that here $R = \rho v a/\eta$. He also found that the friction coefficient $C_f$ ($= 2a(-\partial p/\partial z)/\rho v_m^2$) for turbulent flow was a function only of $R/M$ provided the turbulence was not too vigorous. At high enough values of $R/M$ this simple dependence failed as $C_f$ finally approached its values for zero $M$, which depend simply on $R$. Harris has given an empirical correlation of the results in this latter regime. Murgatroyd's results for the turbulent regime where $C_f$ depends only on $R/M$ may be closely expressed by a generalised version of the universal friction factor formula

$$1/C_f^{\frac{1}{2}} = 12 \log_{10} R(C_f/2)^{\frac{1}{2}}/M - 3 \cdot 2$$

over the range $225 < R/M < 500$. The fact that $C_f$ here depends on $R/M$, which is independent of the channel width, probably indicates that the flow consists of two turbulent boundary layers, isolated from one another by a core of more or less laminar, uniform flow whose width is immaterial. The Hartmann effect makes the mean core velocity uniform and incapable of sustaining turbulence in the face of the direct damping. These details of the situation have not yet been observed experimentally. There is a great need for more experimental work on turbulent magnetohydrodynamic channel flows, not only on flows between parallel planes but even more on flows in circular pipes. The electromagnetic flowmeter is but one of several useful devices in which flows of this type occur.

[*] Hartmann, J. & Lazarus, F. (1937). *Math.-fys. Medd.* 15, no. 7.
[†] Crausse, E. & Poirier, Y. (1957). *C.R. Acad. Sci., Paris,* 244, 2772.
[‡] Murgatroyd, W. (1953). *Phil. Mag.* 44, 1348.
[§] Harris, L. P. (1960). *Hydromagnetic Channel Flows* (Mass. Inst. Tech. Press and Wiley).

# Chapter 4

# OTHER TECHNIQUES OF ELECTRO-
# MAGNETIC FLOW-MEASUREMENT

The previous two chapters have concentrated on exploring the theoretical basis for orthodox transverse-field flowmeters and the less
well-known axial-current flowmeters. In both these types the fluid
passes along a straight duct and the total flow rate or mean velocity is
deduced from measurements of potential differences induced between
two or more electrodes. Two variations on this procedure are considered in this chapter. First, we study electromagnetic velometry,
the measurement of *local* velocities in a fluid stream by means of
observations of induced potentials taken with probes inserted in the
fluid. Then, in the second section of the chapter, we review techniques
of electromagnetic flow-measurement in which some quantity other
than electric potential is observed.

## 4.1. Induction velometry

Fluid dynamicists have always sought to scrutinise the details of
a field of flow in addition to making overall measurements of total
flow rate, etc. Usually the most important quantity to be studied in
detail is the velocity, its magnitude and direction. Various devices
have been developed to this end and used successfully for many years.
Pitot tubes and yawmeters are well established for the measurement
of the magnitude and direction of steady velocities or the mean
velocity in a fluctuating flow. For the measurement of rapidly
fluctuating velocities in gases the hot-wire anemometer has been per-
fected. Its use with liquids is less straightforward and several workers
have tried suitable alternatives. Here we examine the possibility of
using the electromagnetic induction technique. The hope would be to
deduce the fluid velocity from measurements of potential differences
or gradients induced by the fluid's motion in a known, imposed magnetic field. (It will be assumed throughout §4.1 that the magnetic
Reynolds number is so low that the imposed magnetic field is not
significantly perturbed.) It would be expected that the method would
be rapid enough in response to permit the measurement of fluctuating
velocities in turbulence and other unsteady flows, in addition to being

applicable to steady flows. Many of the merits peculiar to induction flowmeters would also characterise induction velometers; an example is the linearity of the calibration.

Like the hot-wire anemometer, an induction velometer would be direction-sensitive. If a measurement of a potential gradient is being made with two, closely spaced probes or electrodes, there are two characteristic directions which define the attitude of the velometer: the local direction of the imposed magnetic field and the direction of the line joining the point electrodes.

The induction technique for finding velocities has been widely exploited by oceanographers. From systematic series of measurements of the p.d. induced by the earth's magnetic field between electrodes towed behind ships, or otherwise, it has proved possible to establish in some detail the pattern of ocean currents in various parts of the world. Several excellent presentations of the theoretical basis for the interpretation of the observations have appeared in the literature and it is not proposed to reproduce that theory here. An extensive list of references to this oceanographic work appears in the Bibliography.

### 4.1.1. *The problem of Ohmic loss*

We shall assume that the d.c. approximation is valid, i.e. that the frequencies of change of the fluid velocity and the imposed magnetic field (if an a.c. magnet is used) are not too high for induced e.m.f.s due to time-varying magnetic fields to be unimportant. This approximation was discussed in §2.1. Then the concept of electric potential is unambiguous.

The aim of induction velometry is to deduce the velocity $\mathbf{v}$ from the induced e.m.f. $\mathbf{v} \times \mathbf{B}$, the magnetic field $\mathbf{B}$ being presumed known. If pairs of closely spaced probes or point electrodes are inserted in the fluid, the signal appearing between them when no current is drawn indicates the potential gradient along the line joining the two points. The induced e.m.f. and the potential gradient unfortunately differ by a term representing the resistance drop associated with any induced currents that are flowing. This is expressed by Ohm's Law,

$$\operatorname{grad} U = \mathbf{v} \times \mathbf{B} - \mathbf{j}/\sigma. \tag{4.1}$$

Thus $\mathbf{v} \times \mathbf{B}$ cannot be directly measured except when induced currents are absent. That situation is the exception rather than the rule. Even if there is no stationary solid conductor in the vicinity, induced currents appear unless $\mathbf{v} \times \mathbf{B}$ is irrotational and can be balanced by the potential gradient. If $\operatorname{curl} \mathbf{v} \times \mathbf{B} \neq 0$, then $\operatorname{curl} \mathbf{j}/\sigma \neq 0$ and $\mathbf{j}$ must occur. Now $\operatorname{curl} \mathbf{v} \times \mathbf{B} = 0$ implies

$$(\mathbf{B}.\operatorname{grad})\,\mathbf{v} = (\mathbf{v}.\operatorname{grad})\,\mathbf{B}, \qquad (4.2)$$

if we take the flow to be incompressible, with $\operatorname{div} \mathbf{v} = 0$.

In many cases of interest the right-hand side of (4.2) vanishes and there is no variation of $\mathbf{B}$ along streamlines. An example is provided by the case of flow in a pipe under a uniform transverse field, the edges of the field being remote. In this case it is impossible for the *left*-hand side of (4.2) to vanish too since the velocity must fall to zero at the walls. Induced currents must occur. The only case of pipe flow where there are no induced currents is in axial-current meters with axisymmetric velocity profiles, where both sides of (4.2) obviously vanish.

The non-vanishing of the right-hand side of (4.2) implies variation of $\mathbf{B}$ along streamlines. An example of this has been encountered earlier in the study of the end-currents at the edges of the field in flowmeters.

The difference between $\operatorname{grad} U$ and $\mathbf{v} \times \mathbf{B}$ may not always be serious. If we consider again the case of pipe flow in a uniform transverse field, it is evident that $(\mathbf{B}.\operatorname{grad})\,\mathbf{v} = 0$ is more nearly satisfied over most of the cross-section for a flat turbulent velocity profile (as was remarked by Kolin* and by Grossman & Charwat†) or a magnetically flattened profile at high $M$, than for a parabolic laminar profile. Indeed, in §3.3.1 the Ohmic term $\mathbf{j}/\sigma$ was justifiably neglected in deriving the value of $S$ for magnetically distorted profiles at high values of $M$. But in general it is not acceptable to assume that measured values of $\operatorname{grad} U$ can be interpreted simply as $\mathbf{v} \times \mathbf{B}$. This is particularly important when measurements of turbulent velocities are being attempted by induction velometry.‡ In their latest paper,§ Grossman et al. are careful to present their results for turbulent pipe flow in terms of fluctuations of electric fields rather than apparent fluctuations of velocity, deduced by ignoring the Ohmic term. This paper describes

\* Kolin, A. (1944). *J. Appl. Phys.* **15**, 150.

† Grossman, L. M. & Charwat, A. F. (1952). *Rev. Sci. Instrum.* **23**, 741.

‡ Borden, A. (Nov. 1950). *David W. Taylor Model Basin Report* 743. Grossman, L. M. & Charwat, A. F. (1952). *Rev. Sci. Instrum.* **23**, 741. Grossman, L. M. & Li, H. (1956). *Univ. of Cal. (Berkeley) Inst. Engng Res. Series* 65, issue 2. Grossman, L. M. *et al.* (1957). *Proc. Amer. Soc. Civ. Engrs* (*J. Hydr. Div.*), **83**, 1394. Grossman, L. M. & Shay, E. A. (1949). *Mech. Engng*, **71**, 744.

§ Grossman, L. M. *et al.* (1957). *Proc. Amer. Soc. Civ. Engrs* (*J. Hydr. Div.*), **83**, 1394.

how the Ohmic drops were themselves scrutinised by making probe measurements of electric fields parallel to the magnetic field, i.e. in a direction in which $\mathbf{v} \times \mathbf{B}$ has no component and the electric field is purely Ohmic in origin.

The question as to how measurements of fluctuating electric fields and their correlations in turbulent shear flow or even in homogeneous turbulence under a uniform imposed magnetic field can be related to the statistical kinematic properties of the turbulence has not yet been answered, but certainly deserves investigation. Fresh urgency has recently been given to the problem of interpreting probe measurements in turbulence in magnetic fields by the prevalence of turbulent plasmas in controlled fusion devices.

Various techniques have been developed for overcoming the problem of the Ohmic term in using electromagnetic induction to measure velocities.

Kolin* suggested the technique of using a tubular, insulating shield round the electrode pair in such a way that the flow passes more or less unimpeded through the tube but the induced currents that circulate round the whole field of flow in the presence of an externally imposed magnetic field are excluded from the fluid between the electrodes. The Ohmic loss is thereby suppressed. Within the velometer tube the non-uniform velocity profile produced by viscosity will engender local induced currents but this will not matter if this velocity profile is axisymmetric. The main difficulty with this device is in ensuring that the fluid motion is not significantly disturbed by the shield. The shield-electrode assembly really needs to be calibrated empirically to allow for the facts that viscosity will depress the mean velocity in the tube below its value in the absence of the tube and that in a tube of limited length there will be end effects, i.e. the induced currents in the main field of flow will to some extent stray into the ends of the tube, causing slight Ohmic loss at the electrodes. A velometer of this type needs to be accurately aligned parallel to the flow and perpendicular to the field and is not well suited to turbulent or unsteady flows.

Kolin† also devised another scheme for eliminating the large-scale induced currents entirely by not having a magnetic field present over the whole cross-section of the pipe but having a concentrated magnetic field merely in the vicinity of the shielded electrodes instead.

---

* Kolin, A. (1944). *J. Appl. Phys.* **15**, 150.

† Kolin, A. (1944). *J. Appl. Phys.* **15**, 150; (1945). *Rev. Sci. Instrum.* **16**, 109.

A sketch of the device appears in fig. 44. The field is provided by a small horseshoe magnet, immersed in the fluid and suitably stream-lined, with the velometer section facing upstream so that the fluid motion there would not be seriously disturbed.

A closely related device is the 'tuyère électrotachymétrique' described by Remeniéras, Hermant and Wolf.* This too is really

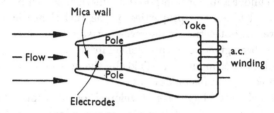

Fig. 44. Kolin velometer (not to scale).

a complete transverse-field flowmeter, enclosed and streamlined so that it can be immersed and traversed in the fluid stream in various hydraulic installations. Figure 45 shows how it takes the form of a

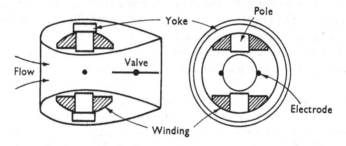

Fig. 45. Tuyère électrotachymétrique. (The valve is a remotely operated butterfly valve for zero setting.)

venturi tube in a streamlined concentric duct with the magnet as-sembly around the venturi and within the duct. The magnetic field and electrodes are located at the throat. Empirical calibration is necessary because the mean velocity at the throat is quite subtly related to the free stream velocity ahead of the device and also end-shorting would be severe. The authors report successful use of the velometer in water down to surprisingly low velocities.

* Hermant, C. & Wolf. (1959). *Houille Blanche,* **14**, 883. Remeniéras, G. & Hermant, C. (1954). *Houille Blanche,* **9**, 732.

Use of a local magnetic field is obviously inescapable in fluid streams of such extent that to provide a uniform magnetic field over the whole field of flow would be difficult and extravagant.

### 4.1.2. *Axisymmetric and other pipe flows*

In the case of axisymmetric, rectilinear flow along circular non-conducting pipes there are other ways of overcoming the problem of Ohmic loss in studies of the velocity distribution under a uniform transverse field, if end-shorting is negligible.

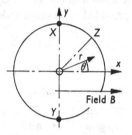

With probes one can measure $U$ or $\partial U/\partial y$ (the potential gradient in the same direction as $\mathbf{v} \times \mathbf{B}$—see fig. 46) at various points in the cross-section and thence deduce the velocity distribution $v(r)$, provided also that the mean velocity $v_m$ is known. This can be found from the potential difference $U_{xr}$ between electrodes $X$ and $Y$ as usual.

To establish the ways in which this can be done we return to the solution of (2.28), which becomes

Fig. 46. Cross-section of circular pipe.

$$\nabla^2 U = Bv' \sin \theta \qquad (4.3)$$

here. The dash denotes differentiation for $r$ and the Laplacian is two-dimensional. If we assume the datum for potential is the centre of the pipe, the solution is of the form $U = Z(r) \sin \theta$ and the boundary condition is $Z'(a) = 0$ since the pipe is non-conducting. $a$ is the pipe radius. Now (4.3) yields

$$Z'' + Z'/r - Z/r^2 = Bv'. \qquad (4.4)$$

We already know that $U_{XY} = 2aBv_m = 2Z(a)$. Integrating (4.4) gives

$$Z' + Z/r = B(v + v_m), \qquad (4.5)$$

in which the last term is the constant of integration, determined at $r = a$ where $Z' = 0$, $Z = aBv_m$ and $v = 0$. Equation (4.5) indicates one way of establishing the distribution $v(r)$ from a potential transverse of the kind first performed by Williams[*] along the line $XY$, $\theta = \frac{1}{2}\pi$. The right-hand side of (4.5) equals $\partial U/\partial y + U/y$ in this case, and knowing $v_m$ we can deduce $v$. The gradient $\partial U/\partial y$ may either be taken from the p.d. between two closely spaced probes or obtained by differentiation of a measured $U$-profile. Alternatively the right-hand side of

---

[*] Williams, E. J. (1930). *Proc. Phys. Soc.*, *Lond.*, **42**, 466.

(4.5) may be written $\dfrac{1}{y}\dfrac{\partial}{\partial y}(Uy)$ and obtained from a graph of the product $Uy$. The accuracy is poor when $y$ is small but fortunately there the velocity only varies slowly in general.

In the above method the determination of the velocity at a point requires knowledge of both the potential and its gradient at the point. Kolin & Reiche[*] pointed out that measurements on the line $OZ$, $\theta = \pi/4$, permit one to find the velocity at a given radius simply from a measurement of $\partial U/\partial y$ with a pair of probes at the same point, provided $v_{\mathrm{m}}$ is known also. The reason for this is that

$$\frac{\partial U}{\partial y} = \left(\sin\theta\,\frac{\partial}{\partial r}+\frac{\cos\theta}{r}\,\frac{\partial}{\partial\theta}\right)Z\sin\theta = Z'\sin^2\theta+\frac{Z}{r}\cos^2\theta$$

$$= \tfrac{1}{2}(Z'+Z/r) = \tfrac{1}{2}B(v+v_{\mathrm{m}}), \quad \text{when} \quad \theta = \pi/4.$$

Thus $\partial U/\partial y$ and $v_{\mathrm{m}}$ suffice to determine $v$ here. Note that it is not sufficient to traverse a single probe along the line $OZ$ because $\partial U/\partial y \neq \partial U/\partial r$ and $\partial U/\partial y$ cannot be found by differentiating $U(r, \pi/4)$.

Kolin & Reiche also pointed out that a possible but more cumbersome alternative is to combine readings of $\partial U/\partial y$ at a given radius at $\theta = 0$ and $\pi/2$ so as to yield values of $v$. When $\theta = 0$, $\partial U/\partial y = Z/r$. Thus, by (4.5),

$$(\partial U/\partial y)_{\theta=0}+(\partial U/\partial y)_{\theta=\pi/2} = B(v+v_{\mathrm{m}}).$$

Instead of using $U_{XY}$ to ascertain $v_{\mathrm{m}}$, it may be established from a reading of a potential gradient by means of a pair of closely spaced electrodes situated anywhere on the periphery of the pipe except at $X$ or $Y$. At the periphery $U = aBv_{\mathrm{m}}\sin\theta$ and the peripheral potential gradient is $Bv_{\mathrm{m}}\cos\theta$. The best place for such a measurement is where $\theta = 0$ or $\pi$ and the signal is largest.

These various measurements all depend critically on the existence of an axisymmetric velocity profile and should not be used unless this is assured by an adequate approach length and the absence of significant magnetohydrodynamic effects. If there is any uncertainty, the axisymmetry of the profile should be checked in some way, e.g. by establishing that $U$ does vary like $\sin\theta$ round the periphery.

The complete velocity distribution cannot be found merely from potential measurements on the periphery. An infinity of velocity profiles can lead to any one peripheral potential distribution.

If the pipe is not circular or the velocity profile is not axisymmetric, the problem of deducing the velocity distribution from the potential

[*] Kolin, A. & Reiche, F. (1954). *J. Appl. Phys.* **25**, 409.

distribution is much more difficult. It is necessary to find the two-dimensional Laplacian $\nabla^2 U$ either by differentiating a complete $U$-profile or by the assembly of five electrodes suggested by Grossman *et al.*\* In this, four electrodes are arranged in a square (of side $h$, say) around a fifth central electrode, the five lying in a cross-section of the pipe. The difference between the sum of the potentials at the outer four electrodes and four times that at the central one, divided by $\frac{1}{2}h^2$, is a good approximation to $\nabla^2 U$ if $h$ is small enough. Then knowing $\nabla^2 U$ we can integrate (2.28) and deduce the velocity

$$v(x, y) = \frac{1}{B} \int_{y_1}^{y} \nabla^2 U \, dy,$$

$y_1$ being one of the two values of $y$ corresponding to the edge of the pipe at each value of $x$ for which the integration is performed. A check is provided by the fact that $v$ should fall to zero again at the other corresponding value of $y$. This method is not restricted to circular or non-conducting pipes. However, it is cumbersome and prone to inaccuracy.

### 4.1.3. *Wall velometers*

At various times several proposals have been made to measure velocities in the vicinity of a solid surface by observation of voltages induced in the presence of a local magnetic field emanating from the wall. The first such suggestion was the scheme of Smith & Slepian† for measuring the speed of ships, the solid surface here being the hull of the ship. Figure 47 shows a two-dimensional view of the device. The flow is perpendicular to the page and the magnet poles are long in this direction to minimise end-shorting. The induced e.m.f. is measured between the electrodes $XY$. A weakness of this device as a ship's log is that the induced signal is dependent more on some mean speed of the water in the boundary layer relative to the hull than on the ship's speed relative to the undisturbed ocean, which will be higher in general.

Essentially similar devices have been more recently used by Guelke & Schoute-Vanneck‡ for measuring the velocity of water near the bed of the sea and by Remeniéras & Hermant§ for measurements in

\* Grossman, L. M. *et al.* (1957). *Proc. Amer. Soc. Civ. Engrs* (*J. Hydr. Div.*), **83**, 1394.

† Smith, C. G. & Slepian, J. (Dec. 1917). U.S. Pat. 1,249,530.

‡ Guelke, R. W. & Schoute-Vanneck, C. A. (1947). *J. Instn Elect. Engrs*, pt. 2, **94**, 71. See also: Longuet-Higgins, M. S. & Barber, N. F. (1946). *Admiralty Res. Lab. Report* RI/102.22/W.

§ Remeniéras, G. & Hermant, C. (1954). *Houille Blanche*, **9**, 732.

hydraulic installations. The same technique has also been attempted
for measurements of ionised gas flow past solid walls. These schemes
involve magnet poles that are short in the direction of flow. The
consequent end-shorting usually necessitates empirical calibration.
The magnets used have been axisymmetric about a line normal to the
wall, as shown in fig. 48 and four electrodes $WXYZ$ have been used in
pairs to detect velocities in the two directions $WZ$ and $XY$. Rotating
the device about $Ox$ until the signal is zero can establish the direction
of flow as being parallel to the line joining the electrodes.

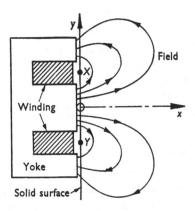

Fig. 47. Smith–Slepian velometer. The fluid flows in the $z$-direction in
the positive-$x$ half-space.

Even in the absence of end-shorting the interpretation of an
observed potential difference is somewhat obscure in these velometers
because of the problem of Ohmic loss. The assertion of Smith &
Slepian that induced currents are absent is erroneous in general. It is
only true if (4.2) holds, the solid wall being assumed non-conducting.
With their device the right-hand side of (4.2) vanishes but the left-
hand side cannot vanish in any real flow near a solid boundary where
velocity gradients must occur. In any case the device really depends
on induced currents for its operation because the p.d. between $X$ and $Y$
in fig. 47 is due solely to the Ohmic drop along the line $XY$, there being
no induced e.m.f. along this line because the fluid velocity is zero
at the solid surface.

*Two-dimensional, Smith–Slepian case.* We may easily predict the
performance of the Smith–Slepian velometer in two extreme cases
provided end-shorting is negligible. The importance of end-shorting
may be inferred approximately from fig. 21, $2c$ being the length of the

94

magnet poles. The two cases are those where the boundary layer is either very thin or very thick in comparison with the characteristic scale $XY$ of the device. Let $XY = 2b$.

If the layer is very thin we may take the velocity $v$ as uniform everywhere with little error and then the induced currents are negligible. In this event

$$U_{XY} = v \int_X^Y B_{xo}\,dy = 2vbB_{xm} \qquad (4.6)$$

in which $B_{xo}$ is the value of $B_x$ at the wall where $x = 0$ and $B_{xm}$ is its mean value over $XY$. The integral may easily be evaluated from

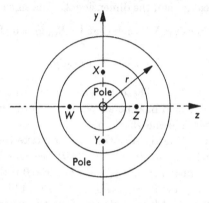

Fig. 48. Axisymmetric wall velometer; plan view. The view in elevation would be the same as fig. 47. The Guelke & Schoute-Vanneck version dispenses with the yoke and uses merely a circular coil.

field-strength measurements on a given meter. To maximise $U_{XY}$, $X$ and $Y$ should be placed at the two points on the line $x = 0$ where $B_x = 0$.

If the layer is very thick, the velocity distribution in the vicinity of the meter will approximate to the shear flow $v = \Omega x$, where $\Omega$ is a constant. The equation governing $U$ is then

$$\nabla^2 U = \operatorname{div} \mathbf{v} \times \mathbf{B} = \mathbf{B} \cdot \operatorname{curl} \mathbf{v} = -\Omega B_y, \qquad (4.7)$$

in which we have used the fact that the magnetic Reynolds number is so low that the field $\mathbf{B}$ in the fluid may be taken as curl-free. The boundary conditions are that $j_x$, $v$ and hence $\partial U/\partial x$ all vanish at $x = 0$ and that $U$ tends to zero at large $x$ and $y$. The magnetic field components $B_x$ and $B_y$ likewise vanish at infinity. The solution of (4.7) in two dimensions is

$$U(x,y) = \tfrac{1}{2}\Omega \left\{ \int_0^y dy \int_y^\infty B_y\,dy + x \int_0^y B_x\,dy \right\},$$

in view of the facts that the magnetic field is symmetrical about $Ox$ and that div $\mathbf{B}$ and curl $\mathbf{B}$ vanish in the fluid. In particular, if the length $XY$ is $2b$, we find that

$$U_{XY} = \Omega \int_0^b dy \int_y^\infty B_{y0} dy = \Omega \left\{ \int_0^b y B_{y0} dy + b \int_b^\infty B_{y0} dy \right\},$$

in which $B_{y0}$ is the value of $B_y$ at the wall where $x = 0$. These integrals can be evaluated from measurements of the field at the wall. Then the Smith–Slepian meter may be used to give a direct measure of the velocity gradient $\Omega$ at the base of a boundary layer provided its thickness is much greater than the dimension $2b$. The signal $U_{XY}$ may be maximised by choosing $XY$ such that $\int_b^\infty B_{y0} dy = 0$. Then

$$U_{XY} = \Omega \int_0^b y B_{y0} dy.$$

*Axisymmetric field case.* These two extreme cases may also be solved when the magnetic field has the axisymmetric form shown in fig. 48, despite the presence of end-shorting.

If the boundary layer is very thin we may take the velocity $v$ as uniform and in the $z$-direction. Now (4.7) becomes $\nabla^2 U = 0$ while the boundary condition is that $j_x = 0$ when $x = 0$ if the wall is non-conducting. Hence $\partial U/\partial x = -v B_y$ there. In addition $U$ and the magnetic field vanish at large distances from the origin. The solution is simply that $\partial U/\partial x = -v B_y$ and $j_x = 0$ everywhere, since $\nabla^2 B_y = 0$. Hence

$$U = v \int_x^\infty B_y dx$$

and

$$U_{XY} = 2v \int_0^\infty B_y dx,$$

the integral being evaluated at $y = b$, $z = 0$. However, because of the axisymmetry of the field, the integral equals $\int_0^\infty B_r dx$, evaluated at $r = b$ and any value of $\theta$, $r$ and $\theta$ being polar co-ordinates in the $y, z$ plane. Then the conservation of magnetic flux leads to the result

$$2\pi b \int_0^\infty B_r dx = 2\pi \int_0^b r B_{x0} dr, \quad \text{evaluated at } x = 0.$$

Thus

$$U_{XY} = \frac{2v}{b} \int_0^b r B_{x0} dr = v b B_{xm}, \qquad (4.8)$$

where $B_{xm}$ is the mean value of $B_{x0}$ at the surface evaluated over the circle of radius $b$ centred at the origin. This result was developed by

Remeniéras & Hermant* who used the unnecessary and unrealistic assumption that the magnetic field is purely in the $x$-direction and does not vary with $x$. The result (4.8) should be compared with (4.6); it is apparent that end-shorting when the field is circular lowers the signal by a factor of two in comparison with the Smith–Slepian case without end-shorting. But it should be noted that $B_{\mathrm{xm}}$ is evaluated on a different basis in the two cases. If $B_{\mathrm{xm}}$ is found from measurements on the meter, (4.8) then provides the calibration law for the meter when the boundary layer is thin. The signal $U_{XY}$ may be maximised by locating $X$ and $Y$ at the two points on the $y$-axis where $B_{\mathrm{xo}} = \frac{1}{2}B_{\mathrm{xm}}$.

We turn next to the situation where the boundary layer is very thick and $v = \Omega x$ in the vicinity of the meter. Again (4.7) applies, even though the field is now axisymmetric about the $x$-axis. The boundary conditions are that $\partial U/\partial x$ vanishes at $x = 0$ and $U$ tends to zero far from the meter. The solution is

$$U(x,y,z) = \tfrac{1}{2}\Omega \left\{ x \int_x^\infty B_{\mathrm{y}}\,dx + \int_x^\infty dx \int_x^\infty B_{\mathrm{y}}\,dx \right\}$$

in view of the facts that $\nabla^2 B_{\mathrm{y}} = 0$ and $B_{\mathrm{y}}$ vanishes at infinity. Thus

$$U_{XY} = \Omega \int_0^\infty dx \int_x^\infty B_{\mathrm{y}}\,dx, \quad \text{evaluated at } y = b, z = 0.$$

It is more useful to express this integral in terms of field intensities at the surface, as was done for the Smith–Slepian meter. When $z = 0$, $B_{\mathrm{y}} = B_{\mathrm{r}}$ and, because $B_{\mathrm{r}}$ is independent of $\theta$, the conservation of magnetic flux leads to the statement

$$\int_x^\infty B_{\mathrm{y}}\,dx = \int_x^\infty B_{\mathrm{r}}\,dx = \frac{1}{b}\int_0^b r B_{\mathrm{x}}\,dr,$$

evaluated at the appropriate value of $x$. Hence

$$U_{XY} = \frac{\Omega}{b}\int_0^\infty dx \int_0^b r B_{\mathrm{x}}\,dr = \frac{\Omega}{b}\int_0^b r\,dr \int_0^\infty B_{\mathrm{x}}\,dx,$$

if we reverse the order of integration. But $\displaystyle\int_0^\infty B_{\mathrm{x}}\,dx = \int_r^\infty B_{\mathrm{ro}}\,dr$, because curl $\mathbf{B} = 0$ and the field vanishes sufficiently rapidly at infinity. $B_{\mathrm{ro}}$ denotes the value of $B_{\mathrm{r}}$ at the surface, $x = 0$. Thus

$$U_{XY} = \frac{\Omega}{b}\int_0^b r\,dr \int_r^\infty B_{\mathrm{ro}}\,dr = \frac{\Omega}{2b}\left\{ \int_0^b r^2 B_{\mathrm{ro}}\,dr + b^2 \int_b^\infty B_{\mathrm{ro}}\,dr \right\}.$$

* Remeniéras, G. & Hermant, C. (1954). *Houille Blanche*, **9**, 732.

This should be compared with the corresponding result for the Smith–Slepian meter. These integrals may be evaluated from measurements of the field at the wall, enabling this axisymmetric configuration also to be used to give a direct indication of the velocity gradient at the base of a boundary layer provided its thickness is much greater than the dimension $2b$. The signal $U_{XY}$ may be maximised by choosing $XY$ such that $b^2 \int_b^\infty B_{ro} dr = \int_0^b r^2 B_{ro} dr$. Then

$$U_{XY} = \frac{\Omega}{b} \int_0^b r^2 B_{ro} dr.$$

*Closing remarks.* It is apparent that induction velometry is not so straightforward as it may appear at first sight because of the complicated relation between potential gradient and velocity. On the other hand, it is very attractive for simple qualitative observations such as determining the direction, sign or degree of steadiness of flow. It is important to remember that induction velometry becomes even more complicated if the magnetic Reynolds number is appreciable and the magnetic field configuration becomes uncertain and itself dependent on the rate of flow. This is only likely to happen with liquid metals. Many of the points made elsewhere in this book in connection with flowmeters apply equally to velometers. The same practical questions such as the choice between a.c. and d.c. operation also arise.

A review of induction velometry would not be complete without mention of Williams's[*] early attempt to explore secondary or swirling flow in a curved pipe by measuring voltages induced by a magnetic field directed *along* the pipe.

### 4.2. Flow-measurement by observations other than induced p.d.

When relatively poorly conducting fluids flow in a magnetic field, the only significant measurable effect that results is the induction of potential differences. The induced currents cannot be measured directly and their effects on the imposed magnetic field or the fluid motion are too feeble to be exploited for flow-measurement. The situation is quite different when we turn to liquid metals, as was seen in ch. 3. There are consequently several alternative ways[†] of measuring the flow of liquid metals by imposing suitable magnetic fields. These

---

[*] Williams, E. J. (1930). *Proc. Phys. Soc., Lond.*, **42**, 466.
[†] Murgatroyd, W. (Oct. 1952). *A.E.R.E. (Harwell) Report* X/R 1053.

are discussed in the subsequent sections of this chapter. We shall distinguish between those relying on dynamic effects and magnetic effects.

### 4.2.1. *Flowmetering by dynamic effects*

The currents induced by the motion of a liquid metal in an imposed magnetic field result in $\mathbf{j} \times \mathbf{B}$ forces acting on the fluid. First consider situations where the liquid is flowing along a pipe under a field that is perpendicular to the motion, due either to external pole-pieces or an axial current. There are several cases to be distinguished.

In regions remote from the edges of the field, currents are known to circulate in planes perpendicular to the motion. If the pipe walls are non-conducting, these currents circulate wholly in the fluid and the associated $\mathbf{j} \times \mathbf{B}$ forces in the fluid produce no net force on the fluid when summed over the whole cross-section. If the pipe walls do conduct, some of these currents enter the walls and then there is a net magnetic force opposing the fluid motion and balanced by a downstream force on the walls. In both these cases there is no perturbation of the magnetic field outside the pipe and no associated force on the source of the imposed field.

At the edges of the field we know that currents circulate more or less in planes perpendicular to the field. There is a resultant $\mathbf{j} \times \mathbf{B}$ force on the fluid opposing its motion and this entails an extra pressure drop in the fluid. It is also associated with a downstream reaction on the source of the field, i.e. the magnet poles in a transverse-field meter, the external conductors in an axial-current meter.

Since we are not now restricted to situations where a measurable p.d. must be generated, configurations in which the induced e.m.f. is entirely short-circuited become of interest. Indeed these tend to be situations where the induced currents and consequent dynamic effects are greatest for a given imposed field intensity and fluid velocity, simply because the induced e.m.f. is wholly devoted to propelling a current. Figure 49 shows an example of this type involving a magnet system which could be applied to an existing pipe without modification. In these cases the current flows wholly in the fluid in planes perpendicular to the motion but there is now a net magnetic force on the fluid, opposing its motion and causing a pressure drop, together with a corresponding downstream reaction on the magnet or coil or both. Note that the behaviour of the axisymmetric device shown in fig. 49 is quite unaffected by wall conductivity or contact resistance.

In any of the above cases the measurable effects which might be exploited for flowmetering are the pressure drop in the fluid, the force on the pipe walls, if conducting, and the force on the magnet, external leads or winding. There is also another dynamic effect, namely, the perturbation of the fluid flow pattern, but this is not attractive for our purposes.

*Pressure-drop flowmeters.* Measuring a pressure drop as an indication of flow rate is already a standard technique, but there may be no strong reason for prefering to induce the drop magnetically rather than by an obstruction, as in orthodox venturi or orifice meters. In any case one of the main reasons for using electromagnetic flowmeters is to avoid having to make pressure measurements in unpleasant fluids like

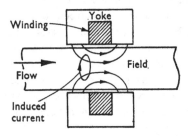

Fig. 49. Axisymmetric force flowmeter. The yoke is not essential.

liquid sodium. The only obvious reason for using a magnetically induced pressure drop is that the relation between drop and flow rate is, to the first order, linear. This gives a simple calibration, unambiguous indication of flow direction and correct recording of the mean value of fluctuating flows. Any non-linearity would be due to serious perturbations of the magnetic field or the velocity field, and due to the fact that some of the pressure loss arises hydraulically.

The production of pressure drops in this fashion has already found a successful application to braking, if not to flowmetering. The rate of natural circulation in liquid sodium coolant loops in nuclear reactors after shut-down has been controlled* by what is essentially a transverse-field flowmeter with a flattened, rectangular channel and an adjustable field. The axisymmetric design shown in fig. 49 is probably better for this purpose, particularly since the pipe may be circular throughout. The fact that there is no braking action on the

---

* Baker, R. S. (Sept. 1958). *U.S.A.E.C. Report* NAA-SR-2986. de Bear, W. S. (June 1959). *Nucleonics,* **17,** 108. Faris, F. E. *et al.* (1958). *Second U.N. Int. Conf. on Peaceful Uses of At. En. (Geneva),* Paper 452, vol. 9, p. 500. Schell, F. N. (Aug. 1953). *U.S.A.E.C. Report* KAPL-M-FNS-6.

pipe axis is unimportant; the fluid readily redistributes its momentum. In §3.3.4 it was seen that the pressure drop across the transverse-field design with a narrow rectangular channel is approximately $0{\cdot}54\sigma B^2 bv_m$, if the edges of the field are abrupt. We are neglecting all but the pressure loss due to end-currents. Faris et al.* gave a similar but more elaborate formula. In the case of the design in fig. 49, the maximum magnitude of $\mathbf{j} \times \mathbf{B}$, the magnetic drag per unit volume, is $\sigma B^2 v$ where $B$ is the maximum value of the *radial* field. The fluid velocity has been taken as uniform. If the axial extent of the field is approximately $2a$, the pipe diameter, and the mean value of (radial field)$^2$ is roughly half the maximum, then the pressure loss is approximately $\sigma B^2 va$ (see Appendix for typical magnitudes).

Another disadvantage of flowmetering by measuring magnetically induced pressure drops is now apparent: the calibration will depend on the fluid property $\sigma$, whose value may be variable and uncertain, particularly in situations where the temperature varies.

*Force flowmeters.* Measuring the force on the pipe walls, if conducting, is precluded by the need to connect these walls to the rest of the piping circuit.

Measuring the force on the magnet or coil, however, is both attractive and practical.† This force varies linearly with flow rate, to the first order, with all the advantages that this implies. A big advantage is that the magnet or coil can be applied externally to a fluid system without any need for breaking or modifying the piping by the addition of electrodes, pressure tappings, etc.

Possible configurations include the transverse-field design, which relies on currents circulating at the edges of the field, and the axisymmetric design in fig. 49. The first alternative has the attraction that the magnet may be applied intact to the pipe whereas the second has to be assembled *in situ* unless the piping can be broken temporarily. A very strong point in favour of the axisymmetric design is that its operation is independent of wall conductivity and contact resistance.

The magnitude of the downstream force on the magnet or coil is readily estimated because it equals the pressure drop due to the eddy currents, already calculated, times the cross-sectional area of the pipe. For the design shown in fig. 49 the force is roughly $\pi\sigma B^2 va^3$ (see Appendix for typical magnitudes). With care this force could be

* Faris, F. E. et al. (1958). *Second U.N. Int. Conf. on Peaceful Uses of At. En. (Geneva)*, Paper 452, vol. 9, p. 500.

† Shercliff, J. A. (Nov. 1952). *A.E.R.E. (Harwell) Report* X/R 1052; (1956). *J. Nucl. Energy*, **3**, 305.

measured to sufficient accuracy. If the magnet yoke was so heavy that it made measurement of the drag force difficult, it could be omitted and the field produced by coils alone, made perhaps of aluminium.

The force flowmeter shares with the pressure-drop meter the disadvantages that its calibration depends on the somewhat imponderable conductivity of the fluid and that its linearity fails as soon as perturbation of either or both of the velocity field and magnetic field becomes significant. The magnitude of these perturbations is measured respectively by the two dimensionless numbers $\sigma B^2 a/\rho v_m$ and $\mu \sigma v_m a$ (the magnetic Reynolds number).

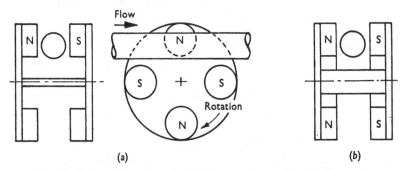

Fig. 50. Rotary flowmeter. In (b) the magnet assembly is axisymmetric with flux return in the shaft.

One suggestion for eliminating the dependence on fluid conductivity is to exploit the idea that if the magnet assembly were to be pulled along freely by the fluid in the absence of significant frictional resistance it would move at some mean synchronous speed, thereby indicating the fluid speed in a manner independent of its conductivity. A practical realisation of this principle necessarily involves rotation of the magnet assembly about some offset axis, because unlimited linear motion is impossible. The design appearing in fig. 50a has been tested* with mercury and the principle of operation demonstrated in practice. With mercury the magnetic forces are so small that friction makes the meter erratic. With sodium it would fare better. The axisymmetric model shown in fig. 50b does not function. The magnetic field must be non-uniform as in fig. 50a. It should be noted that even in the absence of friction the magnet assembly does not rotate synchronously but only at about half that speed. This is because the eddy currents in the fluid oppose the vertical motion of the poles

* Shercliff, J. A. (1957). *A.E.R.E.* (*Harwell*) *Report* X/M 169.

102

even though they promote their horizontal motion when the fluid is moving. These meters will obviously respond poorly to fluctuating flows because of the inertia of the magnet assembly.

### 4.2.2. Flowmetering by magnetic effects: the induced-field flowmeter

Since the motion of liquid metal in an imposed field produces an extra or induced field which is proportional to the flow-rate, to the first order, measurement of this induced field in principle offers yet another technique for flow-measurement. As with the force flowmeter, this method involves an effect which is proportional to the fluid conductivity, which may be uncertain. Indeed, if the velocity is known, the technique can be used to measure the conductivity.* Conductivities of ionised gases in transient flows in shock tubes† have been widely measured by variations on this technique.

The attraction of the induced-field flowmeter is that its indication is directly electrical and no transducer is required. This advantage it shares with orthodox flowmeters in which induced voltages are measured, but in addition the induced-field meter requires no electrical connections to or inside the flow channel. There is no trouble with polarisation of electrodes.

Since a magnetic field is most easily measured by changing it in the presence of a search coil, an a.c. system is to be preferred, the frequency being low enough to avoid skin effect in the liquid metal. This is by no means the only possibility, however. We may take a simple illustrative example. If a d.c. field were provided by a coil or magnet external to the pipe, the configuration of the field outside the pipe would be significantly affected by the fluid motion as soon as the magnetic Reynolds number approached unity. Detecting the changed direction of the field at some suitable spot by a compass needle or otherwise yields a simple, if rudimentary, flowmeter. Alternatively d.c. magnetic fields may be measured by means of movable search coils, Faraday's discs, Hall effect magnetometers, etc. The possibilities are innumerable, and insufficient practical experience has accumulated for any obvious choices to be made.

The earliest schemes for a.c. induced-field flowmeters were those of Lehde & Lang.‡ Their axisymmetric scheme is illustrated in fig. 51. The coils $A$, $B$ and $C$ are enclosed in a streamlined capsule (shown

* Meyer, R. X. (1961). *Second Symp. on the Eng. Asp. of M.H.D.* Philadelphia: Columbia University Press.
† Lin, S. C., Resler, E. L. & Kantrowitz, A. R. (1955). *J. Appl. Phys.* **26**, 95.
‡ Lehde, H. & Lang, W. T. (1948). U.S. Pat. 2,435,043.

dotted in fig. 51) and the fluid motion takes place outside the capsule. There is, however, no reason why the fluid motion should not take place along a pipe inside the coils (the chain-dotted lines in fig. 51). The two coils $A$ and $C$ are energised in series-opposition so as to produce an a.c. field of the form illustrated. In the absence of fluid motion this field is symmetrical and induces no signal in the search coil $B$. As soon as flow occurs the field is dragged downstream and a signal appears in the search coil, proportional to flow rate to the first order.

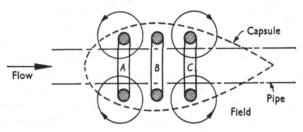

Fig. 51. Lehde & Lang induced-field meter. The capsule and the pipe are alternatives.

A possible improvement would be to make coil $B$ the source of the a.c. field and use $A$ and $C$ in series-opposition as search coils. Optimising the proportions of the coils and their spacing would be a considerable task. It would be possible to use only one coil both to energise the field and to measure the perturbation of the field by the flow. The variation in power factor and impedance of the coil could be interpreted in terms of flow rate, but separating the flow-dependent part of the signal is much less simple than when a separate search coil that yields no signal at no flow is used.

Lehde & Lang were aware that the calibration of the meter would depend on the fluid conductivity and proposed ingenious methods for compensating for this by continuous use of a conductivity cell.

Lehde & Lang also proposed a transverse-field design. Figure 52 shows a side view of the flow channel, looking in the direction of the a.c. field. The field is produced by coils or poles. It is known that the end-currents which occur at the edges of the field weaken the field at the upstream edge and strengthen it at the downstream edge, so producing a flow-dependent signal in the double search coil shown in the figure. Lehde & Lang proposed splitting the flow channel and putting the search coils between the two halves but this is not necessary; it would be sufficient to put the search coils at the side or sides of the channel, next to the poles.

104

Meyer* has described an application of the same principle to measuring the conductivity of ionised air as it flows past a ballistic missile re-entering the earth's atmosphere, the velocity of the air being known. The configuration employed appears in fig. 53. An E-shaped magnet yoke is energised by a.c. windings $A$ and $C$. The symmetry of the field is disrupted by the fluid motion, producing a flow-dependent signal in the search coil $B$. The disturbance of the field is proportional to the magnetic Reynolds number, to the first

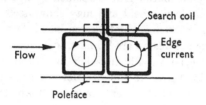

Fig. 52. Arrangement of search coil for transverse induced-field meter.

order, i.e. to the conductivity. Though the conductivity of the ionised air is very low (of order 100 mho/m.), this is compensated for by the fact that the fluid speed is very high in this application.

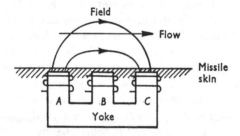

Fig. 53. Conductivity meter described by Meyer.

In the case of induced-field meters of any type in which there is supposed to be no output signal at zero flow, great care is necessary to ensure that this is truly so. Otherwise the genuine signal, which is often quite small, will be lost among stray signals. These designs may be used when $R_m$ is small. However, the other schemes which have been discussed generally demand larger values of $R_m$ simply because the fractional change of the output signal as flow ensues is roughly

* Meyer, R. X. (1961). *Second Symp. on the Eng. Asp. of M.H.D.* Philadelphia: Columbia University Press.

proportional to $R_m$ and would be hard to measure at low $R_m$. Typical magnitudes of $R_m$ appear in the Appendix.

An attraction of induced-field flowmeters is that weak magnetic fields may be used, provided adequate amplifiers are available, thereby saving magnet power and avoiding the danger of dynamic perturbations of the flow.

Another possibility arises in the situation where the magnetic field is produced by an imposed current in the fluid. The perturbation of the magnetic field by the motion is accompanied by a perturbation of the current flow and electric field and may show itself as a variation in the potential difference which has to be applied to the fluid to impose the current. Thus we are led back to induced-voltage meters. A meter of this type was illustrated in fig. 5, but other configurations are possible. Motion in the direction shown leads to e.m.f.s in opposition to the imposed current, i.e. a rise in apparent resistance. Meters of this kind are not limited to a.c. operation. They are only practicable when the magnetic Reynolds number is high, because only then is the fractional change in apparent resistance significant. Another feature of these meters is that they tend to pump or propel the fluid in one direction. In fig. 5 the pumping action is *inwards*, i.e. the pinch effect occurs.

To sum up: induced-field flowmeters and force flowmeters are certainly practicable devices in many circumstances. Their justification is that they usually involve no electrodes and can be installed externally. They cannot compete with the best induced-voltage flowmeters for precision. Only when the magnetic Reynolds number $R_m$ is small does the perturbation of the field and associated force vary linearly with velocity. This occurs because the terms $\mathbf{v} \times \mathbf{B}$ and $\mathbf{j} \times \mathbf{B}$ can be approximated by $\mathbf{v} \times \mathbf{B}_{imposed}$ and $\mathbf{j} \times \mathbf{B}_{imposed}$ when $R_m$ is small and are then linear in $\mathbf{v}$ ($\mathbf{j}$ being linearly dependent on $\mathbf{v}$), but when $R_m$ is larger the non-linear contributions $\mathbf{v} \times \mathbf{B}_{induced}$ and and $\mathbf{j} \times \mathbf{B}_{induced}$ become important. Other non-linear behaviour will occur if the magnetic field is strong enough to perturb the flow pattern. Exact theoretical analysis of induced-field flowmeters and force flowmeters is usually difficult and empirical calibration is inevitable.

# Chapter 5

# AN APPRAISAL OF ELECTROMAGNETIC
# FLOW-MEASUREMENT

This chapter reviews the place that electromagnetic flowmeters do and could occupy in the general field of flow-measurement. Having discussed the case for electromagnetic flowmeters that rely on measuring quantities other than induced p.d. in the second half of the last chapter, here we concentrate on induction meters, i.e. those that do employ measurement of induced p.d. After all, these are the only types which so far have been used at all extensively in practical applications.

The first task is to assess the various qualities of electromagnetic flowmeters that enable them to compete with more orthodox and long established devices. The second is to record the comparative advantages of the different sorts of induction flowmeters, indicating the factors which govern the choice between a.c. and d.c. operation and the choice of geometrical configuration. Many of the pertinent statements have already been made and justified in earlier chapters; our purpose here is to assemble the various points in perspective.

## 5.1. The case for electromagnetic flowmeters

The major claim for the electromagnetic flowmeter is that its output indication—an induced p.d.—is *linearly* dependent on the mean velocity or volumetric flow rate. The extent to which this is actually true is explored in §5.1.1. Taking it as a valid assertion for the moment gives the electromagnetic flowmeter a very strong advantage over most other flowmeters. A linear calibration law implies that the meter indicates the direction of flow unambiguously, unlike a venturi meter, and also that the mean signal in fluctuating flow directly indicates the mean flow rate. The factor relating the signal to the volumetric flow rate depends only on the imposed field strength for a given design and size of meter. The relationship does not involve any properties of the fluid, at least in cases where the pipe walls are non-conducting. The calibration law of such a meter, obtained with

107

one fluid, may be used confidently with another. In contrast, to interpret the indication of orifice meters requires a knowledge of the fluid's density. These meters also show weak dependence on viscosity or Reynolds number.

Another very strong point in favour of electromagnetic flowmeters is that their mode of operation and indication is electric. They may be incorporated in electrical systems for instrumentation and automatic control without the intervention of a transducer. By the same token they are immediately applicable wherever remote indication is essential (although when the fluid is a poor conductor, the length of the leads between the flowmeter and the electrical measuring equipment has to be limited). Chiefly because electrical measurements may be made with high precision, electromagnetic flowmeters, if properly designed and operated, can be made to have calibrations that are linear and reproducible over a long period to $\pm\frac{1}{4}$ per cent, while to achieve more commonplace accuracies of the order of $\pm 1$ per cent is very easy in most situations. Even in difficult applications, where impurities, contact resistance and other imponderables are important, at worst $\pm 5$ per cent accuracy can usually be obtained. Typical of this class are applications to low-purity sodium systems.

Since these meters depend on electric phenomena, they have very rapid response and permit studies of transient flows up to remarkably high frequencies. Granted adequate electrical measuring instruments, the only limits to the response at high frequencies are the onset of fluid self-inductance (skin effect), if the fluid is a liquid metal, or the frequency of excitation of the magnet when an a.c. field is being used, as it probably would be in the case of non-metallic liquids. Arnold* reports the use of a.c. fields at 5000 c/s which permit the detection of transient phenomena up to 1000 c/s. Since the indication does not rely on a mechanical phenomenon, no errors arise from acceleration head as they do when most orthodox meters are applied to unsteady flows. This also means that electromagnetic flowmeters function equally well in such environments as violently accelerating or decelerating missiles or space vehicles.

Most electromagnetic flowmeters present an unimpeded flow passage. Indeed an existing pipe can be turned into a meter by the addition of external electrodes and a suitable magnet if the pipe walls and the fluid are good conductors. Unless the fluid is a liquid metal, the pressure loss associated with the meter is the same as for

* Arnold, J. S. (1951). *Rev. Sci. Instrum.* **22**, 43.

an equivalent length of ordinary piping. The absence of any obstruction to flow is particularly advantageous with fluids like blood or unstable chemicals (where extra disturbance of the fluid or its environment can have undesirable effects on the fluid itself) or with fluids containing large solid bodies which might tend to jam inside rotameters or orifice meters. Nor are there any pressure tappings or other crevices in which solid matter could lodge and which would render cleaning difficult. This is important in sodium circuits where locally or occasionally the metal may solidify. The electromagnetic flowmeter readily accepts slurries and suspensions ranging from sewage to blood and is not affected by severe non-Newtonian behaviour. In circular transverse-field meters, concentric deposition of solid matter on the walls does not affect the calibration if the matter has the same conductivity as the fluid. The meter indicates total volumetric flow rate when used with slurries containing nonconducting suspended solids, provided these are not too large. Having no moving or intricate parts, electromagnetic meters can be built to require no maintenance, a vital consideration in radioactive systems.

It is when the fluid requires special precautions in handling that the case for these meters is strongest. Either the fluid must be protected from contamination (as in the case of blood, pharmaceuticals and foodstuffs) or the fluid must be carefully contained because it is toxic, chemically reactive, inflammable or corrosive (as in the case of radioactive solutions, nuclear reactor coolants or rocket oxidants) or perhaps merely hot. Electromagnetic flowmeters have been employed with liquids up to 1500 °F. The only precaution necessary is to keep the magnet cool enough to stay magnetic at a known intensity. If necessary the field could always be provided by coils alone. There are some situations where the electromagnetic flowmeter has no rival. Probably the best example is its application to the measurement of instantaneous blood flow in the internal arteries of conscious and nearly unimpeded animals.

Most flowmeters are sensitive to upstream disturbances and it is usual to lay down a minimum straight approach length for them. Electromagnetic flowmeters are less demanding in this respect, and can be designed to be completely independent of upstream conditions at the expense of some complication. What exactly constitutes the minimum acceptable upstream length for the simpler designs is still somewhat controversial, but probably never exceeds a few diameters in practice. This means that the meters rarely need to be calibrated *in situ*.

After this eulogy of the electromagnetic flowmeter the reader may be pardoned for wondering why meters of any other kind are still in use! In point of fact, the orthodox rival meters command and will continue to command the major share of the market. The reason is partly conservatism and distrust of 'black boxes' among potential users, but there are many other, more valid reasons. Undoubtedly the major one is expense. The electromagnetic meter is better and more accurate than most industrial applications warrant and so the simpler and cruder orifices, venturis and rotameters are preferred, particularly in view of the high cost of commercially available a.c. electromagnetic meters along with their attendant electronic gear. It should be noted that d.c. flowmeters may be made and operated much more cheaply, the only costly item being a permanent magnet. They are really only suitable for metallic liquids, however.

Apart from cost, the other disadvantage of electromagnetic flowmeters of the types most widely used in practice is that their calibration laws are not accurately predictable. Thus empirical calibration is demanded, but this is easily done because the law is basically linear and independent of the fluid used (unless the walls are conducting). The reason why the performance cannot be predicted is that for cheapness most meters use magnet poles which are short in the direction of flow and therefore end-shorting is severe. It was seen in §2.3 that end-shorting may be estimated easily but not accurately predicted. The problem of designing flowmeters which are truly linear and have predictable calibrations is discussed in §5.2.3. If flowmeter design could be standardised, the corresponding calibration laws could be made generally available and the need for empirical calibration of each new meter would be eliminated.

There are a few other disadvantages of electromagnetic flowmeters. Most types cannot be used where the fluid is flowing in mild steel or other ferromagnetic pipes, because the pipe material shields the fluid. A non-magnetic section has to be inserted in steel pipe circuits. In flow systems where debris or impurity containing iron is present, this may tend to congregate in the region of magnetic field, disrupting and maybe even blocking the meter in the course of time.

Electromagnetic flowmeters may not be used with gases unless they are ionised.

### 5.1.1. *Linearity*

The previous discussion has relied heavily on the assertion that an electromagnetic flowmeter's output signal is stably and linearly proportional to the volumetric rate of flow through it, a statement which occurs very widely without qualification in the literature, but which is not absolutely true in general. Indeed ever since Thürlemann's* theoretical work on circular transverse-field meters, where he showed that the output signal was the same for all axisymmetric velocity profiles for a given flow rate, a certain amount of false confidence has been prevalent.

Consider first the simpler case of a non-metallic fluid conductor. In transverse-field meters non-conducting or insulated walls must be used with such a fluid to avoid total loss of signal, but conducting walls may be used with axial-current meters. There are several reasons why the signal and flow rate may not be mutually proportional. The most important is that the relation between them is usually dependent on the velocity distribution and this varies with Reynolds number, particularly in the presence of upstream effects. In ch. 2 it was shown that the sensitivity $S$ for rectangular non-conducting channels in transverse-field meters can vary by as much as 20 per cent as the velocity profile changes from the laminar to the flat, turbulent form. Thürlemann's result states that this variation does not occur with axisymmetric velocity profiles, laminar or turbulent, in circular transverse-field meters, but even this is only strictly true when the transverse field is absolutely uniform and end-shorting is absent. This is only imperfectly realised in practice.

In electrolytic conductors it is conceivable that further variations in sensitivity could occur because of the variation of the electrical properties of electrolytic flows and boundary layers that various observers have reported.† Contact resistance and other surface effects on electrodes or conducting walls are known to be very variable and can be another source of non-linearity or instability of the calibration of a meter, particularly when the external measuring circuit draws appreciable current. When the walls are conducting, variation of the conductivity of the fluid for any reason also leads to changes in the calibration, as does non-uniformity of the conductivity throughout the fluid.

When a liquid metal is being metered, all the above considerations still apply together with some that are due to the fact that the induced currents are much stronger in liquid metals. Chapter 3 investigated these effects fully and the details will not be reproduced here.

---

* Thürlemann, B. (1941). *Helv. Phys. Acta*, **14**, 383.

† Eskinazi, S. (1958). *Phys. Fluids*, **1**, 161. See also Hogan, M. A. (1923). *Engineering, Lond.*, **115**, 66. Wyatt D. G. (1961). *Phys. Med. Biol.* **5**, 289, 369.

When the flow rate is high the imposed magnetic field in the vicinity of the electrodes can be so severely distorted by the induced currents that the linearity of the meter is affected. This occurs, for instance, in large sodium meters with short pole-faces. On the other hand, at low flow rates the magnetic forces associated with the induced currents can be so strong that they alter the pattern of flow severely enough to disrupt the calibration of the meter. This can occur even in circular transverse-field meters because the magnetic forces make the velocity profile depart from axisymmetry. There are two kinds of magnetic perturbation of the flow to be considered, that occurring at the upstream edge of the field, and that occurring when a steady ultimate state is reached after a sufficient length of uniform field has been traversed. The perturbation that occurs at the edge of the field can be reduced by deliberate fringing. The importance of its effect on the calibration of flowmeters is rather uncertain at the present state of knowledge. The ultimate perturbation is well known from theory and experiment to be quite serious if reached, producing a fall in sensitivity of 7 per cent in circular, transverse-field meters with non-conducting walls when the flow is laminar.

The extent to which these departures from non-linearity actually occur at high and low flow rates depends not only on the velocity but also on the fluid conductivity (and field strength and density also in the case of the velocity profile perturbations). This means that an empirical non-linear calibration that allows for these effects will only be strictly correct for one set of values of the field intensity and fluid properties.

## 5.2. The rival types of electromagnetic induction flowmeters

Many different forms of induction meters are in everyday use and several others have been proposed or tried experimentally. Different applications call for different designs. In each case a suitable compromise has to be reached between conflicting requirements such as cheapness, high accuracy or minimal pressure loss. Some of the factors that govern the design decisions are explored in the following sections.

### 5.2.1. *A.c. versus d.c. operation*

From the earliest days of practical magnetic flow-measuring devices* there have been proposals to use a.c. rather than d.c. fields.

* Smith, C. G. & Slepian, J. (Dec. 1917). U.S. Pat. 1,249, 530.

The main incentive to do this has usually been to avoid polarisation at the electrodes and stray d.c. potentials due to thermoelectric and electrochemical effects. With liquid metals these troubles do not occur at all seriously and d.c. fields are usually employed, although some schools of thought favour a.c. even here. One reason to favour d.c. for liquid metals is the possible occurrence of skin effect. With electrolytic conductors a.c. operation is all but essential. For example, it is standard practice among physiologists to use a.c. flowmeters to measure the flow of blood.

There are other advantages to a.c. as distinct from d.c. operation. An important one is the readiness with which a.c. signals may be amplified. A.c. flowmeters have been used very successfully to measure remarkably small flow rates* and also the flow of very weakly conducting liquids.† Another, rather specialised, point in favour of a.c. is that in axial-current meters‡ in which the current flows in the liquid, it is easier to provide the necessary low-voltage, high-current supply from a single-turn transformer than from a special d.c. source.

In the literature another advantage has been erroneously attributed to a.c. operation. It is not true that the undesirable magneto-hydrodynamic effects do not occur when a.c. fields are used; for a given R.M.S. field strength they are just as important as in the d.c. case.

A.c. operation has many disadvantages that lead one to prefer d.c. whenever there are no stray d.c. signals that must be distinguished from the flow-dependent signal. A major point is that a.c. flowmeters tend to be more complicated and expensive. An a.c. meter usually requires an electromagnet with a laminated yoke and a stabilised power supply whereas a d.c. meter can simply use a permanent magnet. In large meters it is feasible to obtain sufficiently large output signals without the use of a yoke, the field being produced solely by coils. Even more important is the problem of eliminating pick-up, by which is meant the occurrence of stray a.c. signals superimposed on the flow-dependent signal and appearing in the circuit connected to the electrodes. The main source of pick-up is what may be called 'transformer effect', whereby a.c. signals are induced in the circuit by stray alternating fields from the electromagnet. This can never be suppressed entirely and a large effort has been directed into separating the wanted from the stray

* James, W. G. (1951). *Rev. Sci. Instrum.* **22**, 989.
† Lynch, D. R. (Dec. 1959). *Control Engng*, **6**, 122.
‡ Kolin, A. (1956). *J. Appl. Phys.* **27**, 965.

113

signals by many workers. A section in the Bibliography is devoted to the problems of a.c. operation and it would be out of place in this book to go into details. As a rule, separation of the signals is achieved by exploiting the fact that the flow-dependent signal and the pick-up are in quadrature, the first being proportional to the field strength, the second to its time-derivative. This necessitates somewhat elaborate electronic equipment. The situation is actually complicated even further because of phase shifts due to eddy currents in nearby solid and fluid conductors, because of the generation of harmonics through the non-linearity of the material of the magnetic yoke, or because of capacitive pick-up. Another source of trouble can be resonance or beats when the flow contains a slight periodic fluctuation due, for instance, to the use of electromagnetic or mechanical pumps running at or near synchronous speed.

If a.c. operation is decided upon, there remains the choice of frequency. Where 50, 60 or 400 c/s are already available from existing power supplies, these are widely used to avoid the need of special sources, but these frequencies are not necessarily the best. Frequencies ranging from 10 c/s up to 5000 c/s have been used according to reports in the literature. The problem of pick-up is obviously least troublesome when a low frequency is used, but to generate a.c. for the magnet and to amplify the output of the meter are harder at low frequency. There may be other reasons for preferring a high frequency. If it is necessary to study the instantaneous behaviour of transient or pulsating flows, the magnet frequency must clearly be several times higher than the highest frequency of interest occurring in the flow. At the same time there are definite upper limits on the frequency acceptable for the magnet; if the fluid is a good conductor the frequency must not be so high as to cause skin effect, while if the fluid is a poor conductor the frequency $\omega$ must not be so high that dielectric relaxation is not virtually instantaneous. The condition for there to be no skin effect is

$$\omega \ll 1/l^2\mu\sigma$$

and for immediate dielectric relaxation (negligible displacement current) is

$$\omega \ll \sigma/\epsilon.$$

Here $\mu$, $\sigma$ and $\epsilon$ are the fluid's permeability, conductivity and permittivity respectively and $l$ is a representative length scale, such as the pipe diameter (see Appendix for typical magnitudes). Opera-

114

tion of flowmeters under conditions where dielectric effects are important has been discussed by Cushing.*

Other schemes have been developed for circumventing the problem of pick-up in a.c. flowmeters. One method† has been to use a square-wave magnetic field, readings of the induced p.d. being taken only during the quiescent or quasi-d.c. parts of the cycle. More recently a subtler variant‡ on this has been described by the same authors, who use the more easily produced saw-tooth wave-form for the magnetic field. Another school favours the trapezoidal wave-form.§

### 5.2.2. *Transverse-field versus axial-current flowmeters*

Axial-current meters have only been proposed relatively recently and they are not widely in use yet. Nor is it likely that they will oust transverse-field meters from their well-established roles in most cases. However, there are certain situations where axial-current meters do offer special advantages. Most of the advantages and disadvantages attributed to electromagnetic flowmeters in §5.1 apply equally to the two basic types. Here we are concerned only with their merits in relation to one another. Among axial-current meters there is the further choice between having the current flowing in the fluid or in central conductors, insulated from the fluid.

Circular axial-current meters have one clear advantage over transverse-field meters; the pipe walls may be ferromagnetic with impunity. Magnetic walls are quite unacceptable with transverse-field meters. It may be desirable to use magnetic mild steel walls for cheapness or magnetic stainless iron walls to resist corrosion by bismuth. Axial-current meters are very suitable for very hot fluids where there may be difficulty in keeping a magnetic yoke cool enough to stay magnetic.

Another strong point in favour of axial-current meters when the pipe-walls are conducting is that contact resistance, which can be both large and very ill-defined in practice, does not affect the performance of the meter, provided end-shorting is not a serious effect. The reason for this is that there is no tendency for induced currents to circulate between the fluid and the walls, if the velocity profile takes its natural axisymmetric form. This contrasts with the situation prevailing in the transverse-field meter, where such cur-

---

* Cushing, V. (1958). *Rev. Sci. Instrum.* **29**, 692.

† Denison, A. B. & Spencer, M. P. (1956). *Rev. Sci. Instrum.* **27**, 707.

‡ Denison, A. B. & Spencer, M. P. (1960). *Medical Physics* (ed. O. Glasser), vol. 3, p. 178 (Year Book Publ.).

§ Yanof, H. M. (Aug. 1960). *U.S.A.E.C. Report* UCRL-9375.

rents always circulate, permitting contact resistance to exert its baleful effect. The low standard of repeatability of many liquid metal flowmeter tests is undoubtedly due to this. In extreme cases contact resistance causes total loss of signal in transverse-field meters. An axial-current meter may be used with mercury flowing in an un-wetted stainless steel pipe, whereas a transverse-field meter with external electrodes is quite useless for this combination of metals. When walls of high conductivity are being used in order to exploit their 'averaging effect' in axial-current meters, contact resistance, if uniform, even assists this averaging process.

Furthermore, the exact value of the ratio of the conductivities of the fluid and the wall does not affect the performance of axial-current meters (unless the axial current is shared between fluid and wall) whereas it enters critically into the calibration of trans-verse-field meters.

Another way in which the axial-current meter is basically more sound than the transverse-field design is that, if the fluid is a liquid metal, the magnetohydrodynamic forces do not disrupt the natural axisymmetry of the velocity profile, laminar or turbulent, and in fact promote the decay of any asymmetry, the axisymmetric states being those where the sensitivity takes its known, standard value. In contrast, magnetic forces in transverse-field meters tend to distort the velocity distribution to a variable degree, producing variations of sensitivity which are very unpredictable, particularly when the flow is turbulent.

The fact that conducting walls do not tend to short-circuit the signal in axial-current meters (apart from end-shorting) means that metal walls may be used in meters of this type to measure the flow of poor conductors such as water. In these cases the axial current obviously has to flow in a central conductor. In a design such as appears in fig. 26, end-shorting must be minimized by insulating the central conductor assembly from the outer pipe if the fluid is a poor conductor. A significant advantage here is that the effective electrode areas are so large that the output impedance of the device may be low despite the low conductivity of the fluid. To use un-coated metal walls with a non-metallic liquid in orthodox transverse-field meters is futile, on the other hand. It has been pointed out by W. D. Jackson that the situation can be partly saved in this case by using suitably supported interior electrodes $EE$, as in fig. 54, but the resultant signal is very much smaller than in ordinary flow-meter operation and is sensitive to velocity profile.

When the wall conductivity is high the output signal of axial-current meters becomes independent of the velocity profile. Upstream and other disturbances no longer affect the sensitivity as they do in most transverse-field meters. This is strictly true for axial-current meters with the current in the fluid and closely true for annular meters with the current in a central conductor, provided the inner radius of the annulus is not too small.

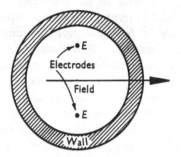

Fig. 54. Internal electrodes in transverse-field meter with highly conducting wall.

Even when the walls are non-conducting, meters with the axial current in the fluid have a sensitivity which is the same for all axi-symmetric velocity profiles, laminar or turbulent, but this virtue they share with circular transverse-field meters.

A final advantage of axial-current meters is that, if one wishes to design a flowmeter whose sensitivity is exactly predictable from theory, the meter must be long enough to eliminate end-shorting, and this is more easily achieved with the axial-current design than with the transverse-field type, for which magnets with long pole-faces soon become prohibitive. A related advantage is that the space normally occupied by the yoke in a transverse-field meter is made available by axial-current meters. There need be no leakage flux at all if the return conductors external to the meter are set co-axially, close to the pipe.

The merits of the axial-current scheme have been set out at some length since it is the less well-known alternative. Some of the points that favour the transverse-field design will now be mentioned briefly.

If it uses a permanent magnet the transverse-field meter does not require a stabilised supply of energising current, and even if it does require a power supply this does not have to be the high-current, low-voltage source demanded by meters with the axial energising

117

current flowing in the liquid. Moreover, it is easier to make the field strengths and consequently the output signals larger in transverse-field meters than in axial-current ones. This is not so important in the larger sizes where shortage of signal is not usually a problem.

Transverse-field meters are hydraulically simpler and much easier to make than annular axial-current meters with central conductors. All designs* for these tend to be somewhat complicated. Even meters with the axial current in the fluid require heavy busbar attachments and also a fragile central electrode on an insulated support. Sometimes transverse-field meters can be created from existing pipework by the mere addition of external electrodes and a magnet. No liquid-tight joints between conductors and insulators are then called for. Only in the case of a highly conducting liquid for which contact resistance could be made low could one use a central-conductor meter of all-welded construction with external electrodes, there being no need for any insulated joints. In all other cases direct contact between the inner and outer walls of the annulus at the current supply points at the ends of the meter would permit ruinous end-shorting to occur.

In short, the case for axial-current meters is probably only overwhelming when circumstances absolutely preclude the transverse-field type. The main merits of the axial-current design are that it can have magnetic walls and can be relatively insensitive to variations of contact resistance and velocity distribution.

This section concludes with a survey of the relative advantages of making the axial current flow in the fluid or in a central conductor.

There are many points in favour of the central-conductor model. The meter where the axial-current flows in the liquid is restricted to use with liquid metals, perhaps even only to the best conductors such as sodium. It demands a low-voltage, high-current supply, which may however be provided by a transformer, integral with the meter*. In central-conductor meters the axial current may be divided between many conductors in series instead. Moreover, the power consumption will obviously be less if the current circulates in copper instead of a liquid metal. Another point is that if the energising current passes through the liquid, some of it will traverse the flow circuit in the wrong direction and interfere with other electrical appliances associated with the circuit.

It has already been remarked that it is desirable to have highly

* Shercliff, J. A. (1959). *J. Nucl. Energy*, Pt. B (*React. Tech.*), **1**, 3.

conducting walls in axial-current meters to make the sensitivity independent of velocity distribution. The central-conductor meter permits this freely, but when the axial current flows in the liquid the sensitivity depends on the somewhat uncertain division of current between the fluid and the walls, if these are conducting. Only if very special circumstances justified the adoption of laminated conducting walls (with peripheral conductivity only) could this dilemma be resolved. Axial current in the walls merely wastes power, moreover.

The central-conductor meter obviates the need for the fragile central electrode that is necessary when the axial current flows in the liquid. When the walls are conducting the electrodes may be situated on the dry sides of the inner and outer walls, and then the fluid is exposed neither to the electrodes nor to their insulation. Since there is no axial potential gradient in the fluid in the central-conductor type, axial positioning of the electrodes is not the crucial consideration that it is when the liquid bears the current.

So far the arguments have all favoured having the energising current in a solid conductor. But this does necessitate flow channels that are hydraulically complicated and difficult to make, and it is necessary to take care that end-shorting is not serious, especially when contact resistance is important and dry (external) electrodes are used. Finally, it should be remembered that the central-conductor meter's sensitivity is to a degree affected by the radial velocity distribution, particularly when the inner radius of the annulus is small.

### 5.2.3. *The design of transverse-field flowmeters*

There are many choices to be made in the design of an electromagnetic flowmeter even if the range of choice has been narrowed down to an a.c. or d.c. transverse-field type that relies on measurement of induced p.d. The cross-sectional shape and materials of the channel must be selected and such details of the magnet or coils as the pole-face length or the degree of fringing determined. If the walls are conducting a choice between wet or dry, interior or exterior, electrodes is available.

There are many factors which govern the various decisions, ranging from cheapness and simplicity to the need for high precision and reproducibility. Sometimes it may be paramount to minimise the pressure loss across the meter. Metallurgical considerations may be dominant. If the fluid is difficult to handle it may be essential

119

to have a meter whose calibration is already known and guaranteed, either from exhaustive previous tests or from reliable theory, so that the need for calibration *in situ* never arises. The presence of inescapable upstream disturbances will sometimes restrict the choice to a design that is insensitive to velocity profile, or the occurrence of uncontrollably variable contact resistance may be a major preoccupation.

This section aims to set forth some of the points that should be borne in mind in choosing a design. Most of the remarks have already appeared in chs. 2 and 3.

Most fluid circuits employ circular piping because it is strong and it is easy to make and join the pipes. For this reason most flowmeters also use a circular section, and, provided the velocity profile is never significantly disturbed from axisymmetry, give very satisfactory service. However fringing of the magnetic field is inevitably severe since the gap has to be large, and it is impossible to predict the calibration accurately unless the pole-faces are made uneconomically long. An empirical calibration must be used.

There is sometimes a strong case for selecting a rectangular channel, provided mechanical strength can be assured. When a calibration that is absolutely insensitive to velocity profile is required, one should use a rectangular channel which is either narrow in the field direction ($b/a \geqslant 5$, say, if 1 per cent precision is required) or has highly conducting walls parallel to the field. The only transverse-field alternative is to use the integrated-voltage designs discussed in ch. 2. Another point in favour of the narrow rectangular channel is that it fits naturally between the poles of a magnet and the magnet is easier to excite than it would be for a circular pipe of the same cross-sectional area. For a given field strength, flow rate and cross-sectional area the circular pipe gives lower output signals also. However, the eddy-current pressure losses across the meter are increased when the depth $2b$ of the channel is increased.

There is no case for using rectangular channels other than the two kinds just mentioned. A non-conducting square or nearly square channel gives a calibration which is even more sensitive to velocity distribution than a circular one.

The selection of materials for constructing the channel demands care. If the fluid is non-metallic a non-conducting or at least an insulated wall is necessary although in rectangular channels the walls parallel to the field may be made conducting to advantage. When the walls are non-conducting, internal, wetted electrodes must be used. The problem of producing leak-proof joints between insulating and conducting materials has to be solved. The electrode area must not be made too small in order to keep the output impedance acceptably low.

When the fluid is a liquid metal it is possible and indeed usual to employ channels with conducting walls. Then external, dry electrodes, not liable to corrosion, may be used. The only reason for still preferring wetted electrodes is that contact resistance does not then so severely deplete the output signal. Whether the electrodes are wet or dry, contact resistance introduces uncertainty into the calibration of a meter with conducting walls. If it proves impossible to eliminate contact resistance in liquid metal systems by achieving high purity or using wetting agents, the only way to obtain a reliable flowmeter calibration is to have non-conducting walls.

Possibly the most vexing question is the choice of pole-face length and the degree of fringing. If the opportunity for empirical calibration is available, the best course is to pack the available magnetic flux into the shortest pole-faces and minimise the fringing, all with a view to making the output signal as large as possible. The fact that the end-shorting cannot be predicted is immaterial.

Circumstances may sometimes demand long pole-faces, however. If the sensitivity of the meter must be exactly predictable from theory the end-shorting must be reduced to a level at which the uncertainty of the available estimates of it is acceptably small. Making $c/b$ larger than 3 is probably good enough for most purposes. If the fluid is a liquid metal and the magnetic Reynolds number can reach values high enough for the magnetic field to be disturbed drastically at its edges, these edges must be made sufficiently remote from the electrodes, i.e. long pole-faces must be used for this reason also. Another danger with liquid metals is variation in sensitivity due to velocity disturbances particularly at the edges of the field when the velocity is low. The best safeguard against this is deliberate enhancement of the fringing of the field by tapering the magnet gap or otherwise. This also has the desirable effect of lessening the pressure drops due to the end-currents.

Another, more complicated, way of eliminating end-shorting and the pressure losses and distortion of the field and the flow due to the end-currents is to install streamwise, insulating vanes or baffles in the fluid just outside the magnetic field. This would be regrettable in an

121

electromagnetic flowmeter, for which a major claim is the fact that it presents no obstacle to the flow.

While it is certainly true that it is possible to build transverse-field meters whose calibration is predictable to high accuracy from theory if the field strength and leading dimensions are known, such flowmeters are not the most desirable from other aspects. They will generally be rectangular ones of the type where $c > b > a$, requiring large magnets, built carefully so as to produce a uniform field, and incurring large pressure drops if the fluid is a liquid metal.

For most purposes a much more useful type of meter would be one whose calibration, though not predictable from theory, was accurately reproducible, and which was cheap and simple. Empirical calibration laws could be established for it to cover a wide range of conditions in the manner that has long been done for orthodox flowmeters.* The data would be presented dimensionlessly so as to apply to different sizes, fluids and field strengths. Though Murgatroyd† several years ago suggested that one or more standard designs of transverse-field flowmeter should be agreed upon and tested exhaustively to yield standardised calibration data, no progress towards this desirable goal can be reported. Each manufacturer uses his own peculiar design of channel and magnet or coils and has his own calibration data. Users who make their own meters must still conduct their own empirical calibrations, against well-established meters such as the venturi, as exhaustively as circumstances warrant. It is high time that electromagnetic flowmeters themselves became well established in this sense, with their own standardised geometrical form and characteristics.

The meter should be as absolute as possible with a calibration that needs a minimum of correction according to circumstances. Because a standard meter must be simple, cheap and robust, the use of a circular pipe is clearly indicated. For a start the design of a meter with non-conducting walls should be standardised. There are some points† in favour of having a venturi contraction in circular meters and putting the electrodes at the throat, but these points are not strong enough to justify departure from the simplicity of a circular pipe of constant diameter. The other main geometrical consideration is the specification of the shape of the magnet poles or coil in the vicinity of the pipe. Here simplicity and ease of manufacture are paramount. The magnet gap should be somewhat

---

\* BS 1042 (1943). Flow measurement.

† Murgatroyd, W. (1952). *A.E.R.E.* (*Harwell*) *Report* X/R 1053.

larger than the pipe diameter in case the fluid is hot. If a permanent magnet is used or if the exciting coils of an electromagnet are not close to the pipe, it is probably sufficient to specify merely the size of the pole faces in relation to the gap and the internal diameter of the pipe. The simplest course would be to take flat, circular pole pieces having a diameter equal to the gap. Compact, commercially available flowmeters usually have the exciting coil and iron yoke arranged to fit closely round the pipe, but even here it should be possible to arrive at a standard, simply defined configuration for the yoke and coil.

The standardised data associated with each design would consist of a basic linear calibration law, involving a single figure typifying the field strength (e.g. the flux density at the centre of the magnet gap). This law would be augmented by whatever corrections proved necessary after the standard designs had been thoroughly tested. Small corrections for variations in the intensity of end-shorting as the ordinary Reynolds number and flow pattern vary may be necessary. Data on the minimum acceptable unimpeded approach length and on pressure losses could also be compiled. The need for extra corrections when the fluid is a liquid metal would have to be explored both at high flow rates (perturbation of the magnetic field) and at low flow rates (perturbation of the flow pattern).

The problem of developing standard flowmeters with conducting walls for liquid metals appears more formidable. It may prove only necessary to augment the data appropriate to the designs with non-conducting walls by providing multiplicatory corrections to allow for the short-circuiting effect of conducting walls in terms of the ratio of the fluid and wall conductivities and the ratio of the internal and external pipe diameters. The main uncertainty will be the effect of contact resistance.

The establishment of standard flowmeter designs and their reproducible characteristics is the main task that deserves attention from experimentalists and flowmeter users. But there are also several other problems about flowmeters that await investigation. On the practical side there exist many schemes for novel types of meters which have not yet been adequately tested. In this category fall axial-current meters, integrated-voltage meters and the various schemes discussed in ch. 4. Meanwhile, there remain many fundamental unsolved questions concerning magnetohydrodynamic phenomena that occur in flowmeters and other magnetohydrodynamic devices. It was pointed out in §3.3.5 that very little experimental information is available on the behaviour of turbulent flows after they enter regions of transverse magnetic field. Another question that needs investigating experimentally is the disturbance of the flow pattern by the end-currents at the edge of the field and the resultant effects on the sensitivity of circular flowmeters. Yet another is the behaviour of swirling flows in a transverse field, although the evidence* is that swirl has no effect on flowmeter performance.

Some controversies still remain even over the design of orthodox transverse-field flowmeters for non-metallic liquids where no magnetohydrodynamic problems arise. These will be settled as more experience with such meters accumulates. No doubt in the hands of the many resourceful people who have developed and will develop designs for meters and their associated measuring equipment, the electromagnetic flowmeter will find application more and more widely and in due course exceed its present high standard of performance and reliability.

* Alden Hydraulic Lab., Worcester (Mass.) Polytechnic (April 1955). *Report on Foxboro meter.*

# Appendix

# TYPICAL MAGNITUDES

### TABLE A1. *Electrical conductivity,* $\sigma$

| | (mho/m) |
|---|---|
| Liquid sodium at 100 °C | $1 \cdot 03 \times 10^7$ |
| NaK eutectic at 100 °C | $2 \cdot 43 \times 10^6$ |
| Mercury at 20 °C | $1 \cdot 05 \times 10^6$ |
| Ionised air near re-entering missile | c. $10^2$ |
| Best electrolyte | c. $10^2$ |
| Sea water | c. 4 |
| Blood | c. 0·7 |
| Tap water | c. $10^{-2}$ |
| Pure water | $4 \times 10^{-6}$ |
| Insulating oil | c. $10^{-11}$ |

### TABLE A2. *Other properties*

| | M.K.S. units | |
|---|---|---|
| | $\rho$ (density) | $\eta$ (viscosity) |
| Liquid sodium at 100 °C | 928 | 0·000686 |
| NaK eutectic at 100 °C | 847 | 0·00047 |
| Mercury at 20 °C | 13,550 | 0·00155 |
| Best electrolyte | c. $10^3$ | c. 0·001 |

### TABLE A3. *Time-varying fields and flows*

*Skin effect.* Angular frequency $\omega$ of field or flow such that skin depth parameter $(\mu\sigma\omega)^{-\frac{1}{2}}$ equals 0·01 m. (1 cm.):

| | Sodium at 100 °C | Mercury at 20 °C | Best electrolyte |
|---|---|---|---|
| $\omega$ (per sec) | 770 | 7600 | $8 \times 10^7$ |

At frequencies much lower than this, skin effect and fluid self-inductance may be neglected if the length-scale is 1 cm. approx.

*Dielectric relaxation.* Angular frequency $\omega$ such that dielectric relaxation number $(\omega\epsilon/\sigma) = 1$:

| | Insulating oil | Pure water | Tap water |
|---|---|---|---|
| $\omega$ (per sec) | 0·45 | 5700 | $1 \cdot 41 \times 10^7$ |
| (taking $\epsilon =$ | $2 \cdot 5\epsilon_0$ | $80\epsilon_0$ | $80\epsilon_0$, |

where $\epsilon_0 = 8 \cdot 854 \times 10^{-12}$, and $\sigma$ from Table A1).

At frequencies much lower than this, dielectric relaxation is virtually instantaneous and displacement current may be neglected.

TABLE A4. *Conducting walls, end-shorting*

Take $a$ or $b = 0.01$ m (1 cm) and assume stainless-steel walls of conductivity $\kappa = 1.4 \times 10^6$ mho/m and thickness $w = 0.001$ m (1 mm).

| Wall conductivity number, $d = w\kappa/a\sigma$ | |
|---|---|
| Sodium at 100 °C | 0.0136 |
| Mercury at 20 °C | 0.133 |
| Best electrolyte | 1400 |

*Contact resistance number*, $\sigma\tau/a$. For unfavourable situations (e.g. mercury and non-wetted stainless steel) this can reach $10^2$ to $10^3$.

*End-shorting number*, $c/bd^{\frac{1}{2}} = (c^2\sigma/wb\kappa)^{\frac{1}{2}}$. Take $c/b = 10$.

| | Sodium at 100 °C | Mercury at 20 °C |
|---|---|---|
| $c/bd^{\frac{1}{2}}$ | 85 | 27 |

TABLE A5. *Magnetohydrodynamic quantities*

Take $B = 0.1$ Wb/m² (1000 gauss), $\mu = 4\pi \times 10^7$ as *in vacuo*, $a = 0.01$ m (1 cm), $v_\mathrm{m} = 1$ m/sec, walls as in Table A4.

| | Sodium at 100 °C | Mercury at 20 °C | Best electrolyte |
|---|---|---|---|
| Reynolds number, $R = \rho v_\mathrm{m} a/\eta$ | 13,500 | 87,000 | 10,000 |
| Magnetic Reynolds number, $R_\mathrm{m} = \mu\sigma v_\mathrm{m} a$ | 0.13 | 0.013 | $1.3 \times 10^{-6}$ |
| Hartmann number, $M = Ba(\sigma/\eta)^{\frac{1}{2}}$ | 123 | 26 | 0.316 |
| (Wall/boundary layer) conductance ratio, $dM$ | 1.67 | 3.45 | no b.l. |
| Edge effect profile distortion number, $\sigma B^2 a/\rho v_\mathrm{m} = M^2/R$ | 1.11 | 0.0078 | $10^{-5}$ |
| Eddy current pressure loss parameter, $\sigma B^2 a v_\mathrm{m}$ | $\begin{cases}1030\\0.15\end{cases}$ | 105 / 0.0152 | $10^{-2}$ Newton/m² / $1.45 \times 10^{-6}$ p.s.i. |
| Force-flowmeter parameter, $\sigma B^2 a^2 v_\mathrm{m}$ | $\begin{cases}0.103\\0.023\end{cases}$ | 0.0105 / 0.0024 | $10^{-6}$ Newton / $2.25 \times 10^{-7}$ lb. |

Note that the last three items vary as $B^2$. Raising $B$ from the modest value of 0.1 Wb/m² soon leads to much bigger effects.

*Ionised air near re-entering missile.* If $\sigma = 10^2$ mho/m, $a = 0.1$ m (10 cm), $v = 0.5 \times 10^4$ m/sec, $R_\mathrm{m} = \mu\sigma v a = 0.063$.

# BIBLIOGRAPHY

1. *Induction flowmeter theory* (see also §§7 and 10)

Arnold, J. S. (1951). An electromagnetic flowmeter for transient flow studies. *Rev. Sci. Instrum.* **22**, 43.

Astley, E. R. (Mar. 1952). Magnetic flowmeter output potentials. *General Electric Report* R 52 GL 52.

Blake, L. R. (1959). Liquid flowmeters. *Brit. Pat.* 802,017 and *Nuclear Engng*, **4**, 238.

Cushing, V. (1958). Induction flowmeter (for use with dielectrics). *Rev. Sci. Instrum.* **29**, 692.

Einhorn, H. D. (1940). Electromagnetic induction in water. *Trans. Roy. Soc. S. Afr.* **28**, 143.

Elrod, H. G. & Fouse, R. R. (1952). An investigation of electromagnetic flowmeters. *Trans. Amer. Soc. Mech. Engrs*, **74**, 589 (formerly *U.S.A.E.C. Report* NEPA 1451, June 1950).

Gessner, U. (1961). Effects of the vessel wall on electromagnetic flow-measurement. *Biophys. J.* **1**, 627.

Gray, W. C. & Astley, E. R. (June 1954). Liquid metal magnetic flow-meters. *Instrum. Soc. Amer. J.* **1**, 15.

Holdaway, H. W. (1957). A note on electromagnetic flowmeters of rectangular cross-section. *Helv. Phys. Acta*, **30**, 85.

Kolin, A. (1945). An alternating field induction flowmeter of high sensitivity. *Rev. Sci. Instrum.* **16**, 109.

Kolin, A. (1952). Improved apparatus and technique for electromagnetic determination of blood flow. *Rev. Sci. Instrum.* **23**, 235.

Kolin, A. (1956). Principle of electromagnetic flowmeter without external magnet. *J. Appl. Phys.* **27**, 965.

Shercliff, J. A. (Nov. 1952). The theory of the d.c. electromagnetic flowmeter for liquid metals. *A.E.R.E.* (*Harwell*) *Report* X/R 1052.

Shercliff, J. A. (1954). Relation between the velocity profile and the sensitivity of electromagnetic flowmeters. *J. Appl. Phys.* **25**, 817.

Shercliff, J. A. (1955). Experiments on the dependence of sensitivity on velocity profile in electromagnetic flowmeters. *J. Sci. Instrum.* **32**, 441.

Shercliff, J. A. (1956). Edge effects in electromagnetic flowmeters. *J. Nucl. Energy*, **3**, 305.

Shercliff, J. A. (1957). Electromagnetic flowmeter without external magnet. *J. Appl. Phys.* **28**, 140.

Shercliff, J. A. (1959). Axial-current electromagnetic flowmeters. *J. Nucl. Energy*, Pt. B (*React. Tech.*), **1**, 3.

Thürlemann, B. (1941). Method for electric speed measurement of fluid. *Helv. Phys. Acta*, **14**, 383. (Also **13** (1940), 343.)

Thürlemann, B. (1955). On the electromagnetic speed measurement of fluid. *Helv. Phys. Acta*, **28**, 483.

127

Williams, E. J. (1930). The induction of e.m.f.s in a moving fluid by a magnetic field and its application to an investigation of the flow of liquids. *Proc. Phys. Soc., Lond.*, **42**, 466.

Wyatt, D. G. (1961). Problems in the measurement of blood flow by magnetic induction. *Phys. Med. Biol.* **5**, 289, 369.

## 2. *Liquid metal induction flowmeters* (theory and experiment)

Affel, R. G. *et al.* (Mar. 1960). Calibration and test of 2 in. and $3\frac{1}{2}$ in. magnetic flowmeters for high temperature NaK service. *U.S.A.E.C. Report* ORNL-2793.

Astley, E. R. (Mar. 1952). Magnetic flowmeter output potentials. *General Electric Report* R 52 GL 42.

Astley, E. R. (Nov. 1954). A report on the calibration of two 8 in. magnetic flowmeters. *U.S.A.E.C. Report* AECU-3171.

Barnes, A. *et al.* (1947, 1952). Electromagnetic flowmeter. *U.S.A.E.C. Reports* ANL-4092 & AECD-3047.

Birch, B. L. (Apr. 1960). Non-nuclear instrumentation for liquid bismuth service. *U.S.A.E.C. Report* BAW-1069.

Bliss, P. (1960). Liquid metal instrumentation practice. *U.S.A.E.C. Report* ORNL-2695, p. 86. (Report of conference, Gatlinburg, Tenn. 1958.)

Carroll, R. M. (Mar. 1953). A simple electromagnetic flowmeter for liquid metals. *U.S.A.E.C. Report* ORNL-1461.

Crofts, T. I. M. (Aug. 1955). The calibration and use of electromagnetic flowmeters in 1 in. stainless steel pipe circuits passing liquid metal (NaK). *U.K.A.E.A. Report* R. & D. B. (W) TN 221.

Duffy, E. J. & Marguin, J. J. (May 1955). Calibration of 8 in. magnetic flowmeter by use of a calibrated orifice. *U.S.A.E.C. Report* KAPL-M-SCT-5.

Elrod, H. G. & Fouse, R. R. (1952). An investigation of electromagnetic flowmeters. *Trans. Amer. Soc. Mech. Engrs*, **74**, 589.

Faris, F. E. *et al.* (1958). Operating experience with the sodium reactor experiment. *Second U.N. Int. Conf. on Peaceful Uses of At. En. (Geneva)*, paper 452, vol. 9, p. 500.

Gasser, E. R. (1957). Operational performance of magnetic flowmeters on a sodium cooled reactor. *U.S.A.E.C. Report* AECU-3853.

Goodman, E. I. & Bakal, R. (Sept. 1950). The design and construction of a test loop for the study of an electromagnetic pump and flowmeter on lithium systems. *U.S.A.E.C. Report* AECU-3622.

Gray, W. C. (Oct. 1951). Magnetic flowmeter calibration results. *U.S.A.E.C. Report* KAPL-613.

Gray, W. C. & Astley, E. R. (June 1954). Liquid metal magnetic flowmeters. *Instrum. Soc. Amer. J.* **1**, 15.

Greenhill, M. & Sabel, C. S. (1956). Electromagnetic pumps and flowmeters. *A.E.R.E. (Harwell) Report* Inf./Bib. 93. Superseded by: Pumps and electromagnetic flowmeters for liquid metals. (May 1959.) *A.E.R.E. (Harwell) Report* Bib. 120.

Hall, W. B. & Crofts, T. I. M. (July 1956). The use of sodium and of sodium-potassium alloy as a heat transfer medium. *J. Brit. Nucl. En. Conf.* **1**, 76.

Huszagh, D. W. (1960). Specialised intrumentation for liquid bismuth loops. *U.S.A.E.C. Report* ORNL-2695, p. 102. (Report of conference, Gatlinburg, Tenn., 1958.)

Kolin, A. (1956). Principle of electromagnetic flowmeter without external magnet. *J. Appl. Phys.* **27**, 965.

Lyon, R. N. (ed.) (1952). *Liquid Metals Handbook.* U.S.O.N.R. Navexos P-733 (Rev.). See also: *Sodium-NaK Supplement,* C. B. Jackson (ed.) (1955).

Murgatroyd, W. (Oct. 1952). The model testing of electromagnetic flowmeters. *A.E.R.E. (Harwell) Report* X/R 1053.

Pfister, C. G. & Dunham, R. J. (Oct. 1957). D.c. magnetic flowmeters for liquid sodium loops. *Nucleonics,* **15**, 122.

Prescott, F. O. (1955). *Reactor Handbook, Engineering,* vol. 2. p. 362, U.S.A.E.C. publ. AECD-3646. McGraw-Hill.

Seban, R. A. *et al.* (Apr. 1949). Flowmetering of molten lead-bismuth eutectic at University of California, Berkeley. *U.S.A.E.C. Report* NP-4452.

Shercliff, J. A. (Nov. 1952). The theory of the d.c. electromagnetic flowmeter for liquid metals. *A.E.R.E. (Harwell) Report* X/R 1052.

Shercliff, J. A. (1956). The flow of conducting fluids in circular pipes under transverse magnetic fields. *J. Fluid Mech.* **1**, 644.

Smith, F. A. (1960). Instruments used in high temperature sodium at Argonne National Laboratory. *U.S.A.E.C. Report* ORNL-2695, p. 67. (Report of conference, Gatlinburg, Tenn., 1958.)

Turner, G. E. (Apr. 1960). The non-linear behaviour of large permanent magnet flowmeters. *U.S.A.E.C. Report* NAA-SR-4544.

Werner, R. C. (Mar. 1955). Liquid metal technology. Final Report. *U.S.A.E.C. Report.* NP-5614.

Yoder, L. W. & Birch, B. L. (Mar. 1959). Calibration tests of a permanent magnet electromagnetic flowmeter for liquid bismuth service. *U.S.A.E.C. Report* BAW-1070.

*Wetting and contact resistance between liquid and solid metal*

Bonilla, C. F. *et al.* (Feb. 1952). Boiling and condensing of liquid metals. *U.S.A.E.C. Reports* NYO-3147 and 3148 (Apr. 1952).

Bonilla, C. F. & Wang, S. J. (Dec. 1952). Interfacial thermal and electrical resistance between stationary mercury and steel. *U.S.A.E.C. Report* NYO-3091.

Droher, J. J. (June 1952). Studies of interfacial effects between mercury and steel. *U.S.A.E.C. Report* ORO-69.

Frost, B. R. T. (1957). The wetting of solids by liquid metals. *Atomics,* **8**, 387.

Kobel, E. (1948). Contact resistance between iron and mercury. *Schweiz. Arch. Angew. Wiss.* **14**, 326.

9                           129                         STE

MacDonald W. C. & Quittenton R. C. (1954). Critical analysis of metal wetting and gas entrainment in heat transfer to molten metals. *Chem. Engng Progr. Symp. Series*, **50**, no. 9, 59.

Milner D. R. (1958). The wetting and spreading of liquid metals on solid metal surfaces. *Brit. Weld. J.* **5**, 90. See also: (1957). *Nature, Lond.,* **180**, 1156.

Vandenburg, L. B. (Feb. 1956). Electrical contact resistance between sodium and stainless steel. *U.S.A.E.C. Report* KAPL-1502.

### 3. *Blood flow measurement*

Albertal, G. *et al.* (1959). Flowmeter for extracorporeal circulation. *Trans. Inst. Radio Engrs,* ME-6, 246.

Clark, J. W. & Randall, J. E. (1949). An electromagnetic blood flow-meter. *Rev. Sci. Instrum.* **20**, 951.

Cobbold, A. F. & Styles, P. (1955). An improved electromagnetic flow-meter. *J. Physiol.* **127**, 1P.

Cooper, T. & Richardson, A. W. (1959). Comparative pulsatile blood flow contours demonstrating the importance of RC output circuit design in electromagnetic blood flowmeters. *Trans. Inst. Radio Engrs,* ME-6, 207.

Cordell, A. R. & Spencer, M. P. (1959). Electromagnetic blood flow measurement in extracorporeal circuits. *Trans. Inst. Radio Engrs,* ME-6, 228.

Denison, A. B. *et al.* (1955). A square-wave electromagnetic flowmeter for application to intact blood vessels. *Circ. Res.* **3**, 39.

Denison, A. B. & Spencer, M. P. (1956). Square wave electromagnetic flowmeter design. *Rev. Sci. Instrum.* **27**, 707.

Denison, A. B. & Spencer, M. P. (1956). Factors involved in intact vessel electromagnetic flow recording. *Fed. Proc.* **15**, 46.

Denison, A. B. & Spencer, M. P. (1960). Circulatory system; methods; magnetic flowmeters. *Medical Physics* (ed. O. Glasser), vol. 3, p. 178 (Year Book Publ.).

Fabre, P. (1932). Use of induced e.m.f.s for recording speed variations of liquid conductors; a new blood flow measurement without palette. *C.R. Acad. Sci., Paris,* **194**, 1097.

Feder, W. & Bay, E. B. (1959). The d.c. electromagnetic flowmeter and its application to blood flow measurement in unopened vessels. *Trans. Inst. Radio Engrs,* ME-6, 240, 250.

Ferguson, D. J. & Wells, H. S. (1959). Frequencies in pulsatile flow and response of magnetic meter. *Circ. Res.* **7**, 336.

Flores, A. *et al.* (1956). The determination of blood flow through intact human arteries. *Surg. Forum,* **6**, 224.

Giori, F. A. (1960). A blood flowmeter for use in coronary heart disease research. *Trans. Inst. Radio Engrs,* ME-7, 211.

Inouye, A. & Kuga, H. (1954). On the applicability of electromagnetic flowmeters for the measurement of blood flow rate. *Jap. J. Physiol.* **4**, 205.

Inouye, A. *et al.* (1955). New method for recording pressure/flow diagram applicable to peripheral blood vessels of animals. *Jap. J. Physiol.* 5, 236.

Inouye, A. & Kosaka, H. (1959). A study of flow patterns in carotid and femoral arteries of rabbits and dogs with the electromagnetic flowmeter. *J. Physiol.* 147, 209.

Jochim, K. E. (1948). Electromagnetic flowmeter. *Meth. Med. Res.* 1, 108.

Jochim, K. E. (1950). Circulatory system; methods; electromagnetic flowmeters. *Medical Physics* (ed. O. Glasser), vol. 2, p. 225 (Year Book Publ.).

Katz, L. N. & Jochim, K. E. (1944). Electromagnetic flowmeter. *Medical Physics* (ed. O. Glasser), vol. 1, p. 377 (Year Book Publ.).

Katz, L. N. & Kolin, A. (1938). The flow of blood in the carotid artery of the dog under various circumstances as determined with an electromagnetic flowmeter. *Amer. J. Physiol.* 122, 788.

Kolin, A. (1936). An electromagnetic flowmeter. The principle of the method and its application to blood flow measurement. *Proc. Soc. Exp. Biol., N.Y.*, 35, 53.

Kolin, A. (1937). An electromagnetic recording flowmeter. *Amer. J. Physiol.* 119, 355.

Kolin, A. & Katz, L. N. (1937). Observation of the instantaneous speed of blood with an electromagnetic flowmeter. *Ann. Physiol. Physicochim. Biol.* 13, 1022.

Kolin, A. (1941). An A.C. induction flowmeter for measurement of blood flow in intact blood vessels. *Proc. Soc. Exp. Biol., N.Y.*, 46, 235.

Kolin, A. *et al.* (1941). Electromagnetic measurement of blood flow and sphygmomanometry in the intact animal. *Proc. Soc. Exp. Biol., N.Y.*, 47, 324.

Kolin, A. (1952). Improved apparatus and technique for electromagnetic determination of blood flow. *Rev. Sci. Instrum.* 23, 235.

Kolin, A. *et al.* (1957). Electromagnetic determination of regional blood flow in unanaesthetised animals. *Proc. Nat. Acad. Sci., Wash.*, 43, 527.

Kolin, A. (1959). Electromagnetic blood flowmeters. *Science*, 130, 1088.

Kolin, A. & Kado, R. T. (1959). Miniaturisation of the electromagnetic blood flowmeter and its use for recording circulatory responses of conscious animals to sensory stimuli. *Proc. Nat. Acad. Sci., Wash.*, 45, 1312.

Kolin, A. (1960). Circulatory system; methods; blood flow determination by the electromagnetic method. *Medical Physics* (ed. O. Glasser), vol. 3, p. 141 (Year Book Publ.).

McDonald, D. A. (1955). The relation of pulsatile pressure to flow in arteries. *J. Physiol.* 127, 533.

McDonald, D. A. (1960). *Blood Flow in Arteries.* (Arnold.)

Olmsted, F. (1959). Measurement of cardiac output in unrestrained dogs by an implanted electromagnetic meter. *Trans. Inst. Radio Engrs*, ME-6, 210.

131

Randall, J. E. & Stacy, R. W. (1956). Pulsatile and steady pressure-flow relations in the vascular bed of the hind leg of the dog. *Amer. J. Physiol.* **185**, 351.

Richards, T. G. & Williams, T. D. (1953). Velocity change in carotid and femoral arteries of dogs during cardiac cycles. *J. Physiol.* **120**, 257.

Richardson, A. W. *et al.* (1949). A newly developed electromagnetic flowmeter. *J. Lab. Clin. Med.* **34**, 1706.

Richardson, A. W. *et al.* (1952). A newly modified electromagnetic flowmeter capable of high fidelity flow registration. *Circulation*, **5**, 430.

Shirer, H. W. *et. al.* (1959). A magnetic flowmeter for recording cardiac output. *Proc. Inst. Radio Engrs*, **47**, 1901; *Trans. Inst. Radio Engrs*, ME-6, 232.

Spencer, M. P. *et al.* (1955). Electromagnetic measurement of blood flow through intact human arteries. *Amer. J. Med.* **19**, 153.

Spencer, M. P. & Denison, A. B. (1956). The aortic flow pulse as related to differential pressure. *Circ. Res.* **4**, 476.

Spencer, M. P. *et al.* (1958). Continuous measurement of cardiac output in conscious dogs by means of an indwelling magnet and the square wave magnetic flowmeter. *Fed. Proc.* **17**, 154.

Spencer, M. P. *et al.* (1959). Measurement of blood flow in experimental animals by means of implanted electromagnetic probes. *Fed. Proc.* **18**, 150.

Spencer, M. P. & Denison, A. B. (1959–60). Square wave electromagnetic flowmeter. Theory of operation and design of magnetic probes for clinical and experimental applications. *Trans. Inst. Radio Engrs*, ME-6, 220; *Meth. Med. Res.* **8**, 321.

Wetterer, E. (1937). A new method for measuring the rate of blood circulation in an unopened vessel. *Z. Biol.* **98**, 26.

Wetterer, E. (1938). The induction tachygraph. *Z. Biol.* **99**, 158, 320.

Wetterer, E. & Deppe, B. (1940). *Z. Biol.* **100**, 205.

Womersley, J. R. (1958). Oscillatory flow in arteries. The reflection of the pulse wave at junctions and rigid inserts in the arterial system. *Phys. Med. Biol.* **2**, 313.

Wyatt, D. G. (1961). Problems in the measurement of blood flow by magnetic induction. *Phys. Med. Biol.* **5**, 289, 369.

Yanof, H. M. & Salz, P. (June 1960). A trapezoidal-wave electromagnetic blood flowmeter. *U.S.A.E.C. Report* UCRL-9204.

Yanof, H. M. (Aug. 1960). A new trapezoidal-wave electromagnetic blood flowmeter and its application to the study of blood flow in the dog. *U.S.A.E.C. Report* UCRL-9375.

4. *Some other specialised applications of induction flowmeters*

Arnold, J. S. (1951). An electromagnetic flowmeter for transient flow studies. (Waterhammer, etc.). *Rev. Sci. Instrum.* **22**, 43.

Babcock, R. H. (1956–57). Magnetic flowmeter. *Publ. Wks, N.Y.,* **87**, 93. *Wat. & Sewage Wks, N.Y.,* **104**, 380.

Balls, B. W. & Brown, K. J. (1959). Magnetic flowmeter (slurries, pulp). *Trans. Soc. Instrum. Tech.* **11**, 123.

Drinker, H. (Apr. 1956). The magnetic flowmeter (pulp stock). *J. Tech. Ass. Pulp Paper Ind.* **39**, 185 A.

Grundy, L. A. (1955). Electromagnetic flowmeter in latex service. *Industr. Engng Chem.* **47**, 2445.

Jaffe, L. *et al.* (Mar. 1951). An electromagnetic flowmeter for rocket research (nitric acid oxidant). *NACA Report* RM E 50 L 12.

James, W. G. (1951). An induction flowmeter suitable for radioactive liquids. *Rev. Sci. Instrum.* **22**, 989 (formerly *U.S.A.E.C. Report* ORNL-1013 (June 1951)).

Kellex Corp. (1949–51). Magnetic induction flowmeter development (radio-chemicals). *U.S.A.E.C. Reports* KLX-1037 (June 1949), 1312 (Aug. 1950), 1321 (Dec. 1950), 1334 (Feb. 1951), and 1347 (June 1951).

Morris, A. J. & Chadwick, J. H. (1951). An electromagnetic induction method of measuring oscillatory fluid flow (hydraulic stabiliser). *Trans. Amer. Inst. Elect. Engrs*, **70**, 346; *Elect. Engng, N.Y.*, **70**, 529.

Remeniéras, G. & Hermant, C. (1954). Electromagnetic measurement of speed in liquids (hydraulics). *Houille Blanche*, **9**, 732.

Richardson, L. M. (1956). Automatic flow control improves filter performance; a magnetic flowmeter in a brewery. *Food Engng*, **28**, 72.

Vitro Corp. (1952–53). Magnetic induction flowmeter development (radiochemicals). *U.S.A.E.C. Reports* KLX-1361 (Oct. 1952) and 1391 (July 1953).

5. *A.c. flowmeter techniques; circuitry; magnet design*

Abel, F. L. (1959). Chopper-operated electromagnetic flowmeter. *Trans. Inst. Radio Engrs*, ME-6, 216.

Ageikin, D. I. & Desova, A. A. (1956). An electromagnetic flowmeter. *Avtomatika i Telemekhanika*, **17**, 1123.

Arnold, J. S. (1951). An electromagnetic flowmeter for transient flow studies. *Rev. Sci. Instrum.* **22**, 43.

Balls, B. W. & Brown, K. (1959). Magnetic flowmeter. *Trans. Soc. Instrum. Tech.* **11**, 123.

Barnes, C. W. (1960). A new method for obtaining flow signals from the electromagnetic flowmeter. *Naturwissenschaften*, **47**, 56.

Clark, J. W. & Randall, J. E. (1949). An electromagnetic blood flowmeter. *Rev. Sci. Instrum.* **20**, 951.

Cooper, T. & Richardson, A. W. (1959). Comparative pulsatile blood flow contours demonstrating the importance of RC output circuit design in electromagnetic blood flowmeters. *Trans. Inst. Radio Engrs*, ME-6, 207.

Denison, A. B. & Spencer, M. P. (1956). Square wave electromagnetic flowmeter design. *Rev. Sci. Instrum.* **27**, 707.

Denison, A. B. & Spencer, M. P. (1960). Circulatory system; methods; magnetic flowmeters (sawtooth wave-form). *Medical Physics* (ed. O. Glasser), vol. 3, p. 178 (Year Book Publ.).

Einhorn, H. D. (1940). Electromagnetic induction in water. *Trans. Roy. Soc. S. Afr.* **28**, 143.

Elrod, H. G. & Fouse, R. R. (1952). An investigation of electromagnetic flowmeters. *Trans. Amer. Soc. Mech. Engrs,* **74**, 589.

Feder, W. & Bay, E. B. (1959). The d.c. electromagnetic flowmeter and its application to blood flow measurement in unopened vessels (electrodes). *Trans. Inst. Radio Engrs,* ME-6, 240, 250.

Ferguson, D. J. & Wells, H. S. (1959). Frequencies in pulsatile flow and response of magnetic meter. *Circ. Res.* **7**, 336.

Guelke, R. W. & Schoute-Vanneck, C. A. (1947). The measurement of sea water velocities by electromagnetic induction. *J. Instn Elect. Engrs,* **94**, pt. 2, 71.

Head, V. P. (1959). Electromagnetic flowmeter primary elements. *Trans. Amer. Soc. Mech. Engrs,* **81**, Series D, 660.

Hogg, W. R. *et al.* (1952). Electronic circuit problems in electromagnetic flow measurement. *Proc. Nat. Electron. Conf.* **8**, 127.

Hutcheon, I. C. (Apr. 1960). Some problems of magnetic flow measurement. *Instrum. Engng,* **3**, 1.

Jaffe, L. *et al.* (Mar. 1951). An electromagnetic flowmeter for rocket research. *NACA Report* RM E 50 L 12.

James, W. G. (1951). An induction flowmeter design suitable for radioactive liquids (very low flow rates). *Rev. Sci. Instrum.* **22**, 989.

James, W. G. (1952). An A.C. induction flowmeter. *Instruments,* **25**, 473.

Jochim, K. E. (1939). Some improvements on the electromagnetic flowmeter. *Amer. J. Physiol.* **126**, 547.

Jochim, K. E. (1948). Electromagnetic flowmeter. *Meth. Med. Res.* **1**, 108.

Jochim, K. E. (1950). Circulatory system; methods; electromagnetic flowmeter. *Medical Physics* (ed. O. Glasser), vol. 2, p. 225 (Year Book Publ.).

Katz, L. N. & Jochim, K. E. (1944). Electromagnetic flowmeters. *Medical Physics* (ed. O. Glasser), vol. 1, p. 377 (Year Book Publ.).

Katz, L. N. & Kolin, A. (1938). Flow of blood in carotid artery of the dog under various circumstances as determined with electromagnetic flowmeter. *Amer. J. Physiol.* **122**, 788.

Kolin, A. (1941). An a.c. induction flowmeter for measurement of blood flow in intact blood vessels. *Proc. Soc. Exp. Biol., N.Y.,* **46**, 235.

Kolin, A. (1941). A variable-phase transformer and its use as an A.C. interference eliminator. *Rev. Sci. Instrum.* **12**, 555.

Kolin, A. (1945). An alternating field induction flowmeter of high sensitivity. *Rev. Sci. Instrum.* **16**, 109.

Kolin, A. (1953). A method for adjustment of the zero setting of an electromagnetic flowmeter without interruption of the flow. *Rev. Sci. Instrum.* **24**, 178.

Kolin, A. *et al.* (1958). Electromagnetic blood flowmeter yielding a base line without interruption of flow. *Proc. Soc. Exp. Biol., N.Y.,* **98**, 550.

Kolin, A. & Kado, R. T. (1959). Miniaturisation of the electromagnetic blood flowmeter and its use for recording circulatory responses of conscious animals to sensory stimuli. *Proc. Nat. Acad. Sci., Wash.*, **45**, 1312.

Kolin, A. & Kado, R. T. (1960). Simple photoelectric demodulator. *J. Sci. Instrum.* **37**, 107.

Kolin, A. (1960). Circulatory system; methods; blood flow determination by the electromagnetic method. *Medical Physics* (ed. O. Glasser), vol. 3, p. 141 (Year Book Publ.).

Lynch, D. R. (Dec. 1959). A low-conductivity magnetic flowmeter. *Control Engng*, **6**, 122.

Morris, A. J. & Chadwick, J. H. (1951). An electromagnetic induction method of measuring oscillatory fluid flow. *Trans. Amer. Inst. Elect. Engrs*, **70**, 346.

Nikitin, B. I. (1956). Liquid flow measurement by the electromagnetic method. *Priborostroenie*, **7**, 13.

Olmsted, F. (1959). Measurement of cardiac output in unrestrained dogs by an implanted electromagnetic meter. *Trans. Inst. Radio Engrs*, ME-6, 210.

Olmsted, F. & Aldrich, F. D. (1961). Improved electromagnetic flowmeter; phase detection, a new principle. *J. Appl. Physiol.* **16**, 197.

Remeniéras, G. & Hermant, C. (1954). Electromagnetic measurement of speed in liquids. *Houille Blanche*, **9**, 732.

Rolff, J. J. P. (1960). Magnetic flowmeters. *Arch. Tech. Messen.* no. 297, p. 197.

Shirer, H. W. *et al.* (1959). A magnetic flowmeter for recording cardiac output (square wave). *Proc. Inst. Radio Engrs*, **47**, 1901.

Spencer, M. P. & Denison, A. B. (1960). Square wave electromagnetic flowmeter for surgical and experimental applications. *Meth. Med. Res.* **8**, 321.

Vitro Corp. (July 1953). Magnetic induction flowmeter development. *U.S.A.E.C. Report* KLX-1391.

Westerten, A. *et al.* (1959). Gated sine-wave electromagnetic flowmeter. *Trans. Inst. Radio Engrs*, ME-6, 213.

Wyatt, D. G. (1961). Problems in the measurement of blood flow by magnetic induction. *Phys. Med. Biol.* **5**, 289, 369.

Yanof, H. M. & Salz, P. (June 1960). A trapezoidal-wave electromagnetic blood flowmeter. *U.S.A.E.C. Report* UCRL-9204.

Yanof, H. M. (Aug. 1960). A new trapezoidal-wave electromagnetic blood flowmeter and its application to the study of blood flow in the dog. *U.S.A.E.C. Report* UCRL-9375.

6. *Electromagnetic velometry*

Borden, A. (Nov. 1950). The development of a turbulence indicator for liquids utilising magnetic induction. *David W. Taylor Model Basin Report* 743.

Grossman, L. M. & Shay, E. A. (1949). Turbulent velocity measurements. *Mech. Engng*, **71**, 744.

Grossman, L. M. & Charwat, A. F. (1952). The measurement of turbulent velocity fluctuations by the method of electromagnetic induction. *Rev. Sci. Instrum.* **23**, 741.

Grossman, L. M. & Li, H. (Feb. 1956). Turbulence investigations in liquid shear flow by the method of electromagnetic induction. *Univ. of Cal. (Berkeley) Inst. Engng Res. Series* **65**, Issue 2.

Grossman, L. M. *et al.* (1957). Turbulence in civil engineering. Investigation of liquid shear flow by electromagnetic induction. *Proc. Amer. Soc. Civ. Engrs (J. Hydr. Div.)*, **83**, 1394.

Guelke, R. W. & Schoute-Vanneck, C. A. (1947). The measurement of sea water velocities by electromagnetic induction. *J. Instn Elect. Engrs*, **94**, pt. 2, 71.

Hermant, C. & Wolf, M. (1959). Some practical applications of the electromagnetic nozzle for the measurement of low velocities. *Houille Blanche*, **14**, 883.

Kolin, A. (1943). Electromagnetic method for the determination of velocity distribution in fluid flow. *Phys. Rev.* **63**, 218.

Kolin, A. (1944). Electromagnetic velometry. I. A method for the determination of fluid velocity in space and time. *J. Appl. Phys.* **15**, 150.

Kolin, A. (1945). An alternating field induction flowmeter of high sensitivity. *Rev. Sci. Instrum.* **16**, 109.

Kolin, A. & Reiche, F. (1954). Electromagnetic velometry. II. Elimination of the effects of induced currents in explorations of the velocity distribution in axially symmetric flow. *J. Appl. Phys.* **25**, 409.

Longuet-Higgins, M. S. & Barber, N. F. (1946). The measurement of water velocities by electromagnetic induction. An electrode flowmeter. *Admiralty Res. Lab. Report* R 1/102.22/W.

Remeniéras, G. & Hermant, C. (1954). Electromagnetic measurement of speed in liquids. *Houille Blanche*, **9**, 732.

Smith, C. G. & Slepian, J. (Dec. 1917). Electromagnetic ship's log. U.S. Pat. 1,249,530.

Williams, E. J. (1930). The induction of e.m.f.s in a moving liquid by a magnetic field and its application to an investigation of the flow of liquids. *Proc. Phys. Soc., Lond.*, **42**, 466.


7. *Oceanography, electric currents in rivers, submarine cables, etc.*

Barber, N. F. & Longuet-Higgins, M. S. (1948). Water movements and earth currents; electric and magnetic effects. *Nature, Lond.*, **161**, 192.

Bernard, M. (1938). Observations of the earth current in a submarine cable. *Onde Élect.* **17**, 465.

Bowden, K. F. (1953). Measurement of wind currents in the sea by the method of towed electrodes. *Nature, Lond.*, **171**, 735.

Buchholz, H. (1958). Electric currents induced by the earth's magnetic field in the moving water of a river (theory). *Arch. Elektrotech.* **44**, 12.

Cagniard, L. (1957). On the theory of the electromagnetic method of measuring ocean currents. *Ann. Géophys.* **13**, 155.

Cherry, D. W. & Stovold, A. T. (1946). Earth currents in short submarine cables. *Nature, Lond.*, **157**, 766.

Dechevrens, M. (1923). (Tidally induced currents.) *Rev. Quest. Sci.* **83**, 302.

Dresing, H. (1881). (Tidally induced currents in cables.) *J. Soc. Tel. Engrs*, **10**, 71.

Ezoe, T. & Suzuki, K. (1956). On the faults of submarine cables by electrolytic corrosion. *J. Inst. Elect. Engrs, Japan*, **76**, 609.

Faraday, M. (1832). Experimental researches in electricity. *Phil. Trans.* **15**, 175. *Diary*, vol. 1, p. 409 (publ. G. Bell and Sons, London, 1932). *Faraday's Experimental Researches in Electricity*, vol. 1. p. 55 (publ. Taylor and Francis, London, 1839).

Guelke, R. W. & Schoute-Vanneck, C. A. (1947). The measurement of sea water velocities by electromagnetic induction. *J. Instn Elect. Engrs*, **94**, pt. 2, 71.

Kiyono, T. & Ezoe, T. (1957, 1959). On the electric field due to tides, I, II, and III. *Mem. Fac. Engng Koyoto Univ.* **19**, 255 (I—cable corrosion); **21**, 15, 170 (II, III—oceanography).

Le Grand, Y. (1956). Electric currents and potential differences in the sea (theory). *Bull. Inf. C.O.E.C. VIII*, no. 1, 11.

Longuet-Higgins, M. S. (1947). The electric field induced in a channel of moving water. *Admiralty Res. Lab. Report* R 2/102.22/W.

Longuet-Higgins, M. S. (1949). The electric and magnetic effects of tidal streams. *Mon. Not. R. Astr. Soc. Geophys. Suppl.* **5**, 285.

Longuet-Higgins, M. S. *et al.* (1954). The electric field induced by ocean currents and waves, with application to the method of towed electrodes. *Pap. Phys. Oceanogr. Met.* (M.I.T. and Woods Hole), **13**, no. 1.

Malkus, W. V. R. & Stern, M. E. (1952). Determination of ocean transports and velocities by electromagnetic effects. *J. Mar. Res.* **11**, 97.

Martin, J. (1956). Use of electric current-meter with towed electrodes. *Bull. Inf. C.O.E.C. VIII*, no. 8, 355, 465.

Moroshkin, K. V. (1957). Experimental work with electromagnetic measuring instrument for currents in the open sea. *Trud. Inst. Okeanol. Akad. Nauk SSSR*, **25**, 62.

Regnart, H. C. (1930). Generation of electric currents by water moving in a magnetic field. *Proc. Univ. Durham Phil. Soc.* **8**, 291.

Rikitake, T. (1960). Electromagnetic induction in a hemi-spherical ocean. *J. Geomagn. Geolect., Kyoto*, **11**, 65.

Saunders, H. (1881). (Tidally induced currents in cables.) *J. Soc. Tel. Engrs.* **10**, 46.

Stommel, H. M. (1948). The theory of the electric field induced in deep ocean currents. *J. Mar. Res.* **7**, 386.

Stommel, H. M. (1956). Electrical data from cable may aid hurricane prediction. *Western Union Tech. Rev.* **10**, 15.

Thürlemann, B. (1955). On the electromagnetic speed-measurement of fluid. *Helv. Phys. Acta*, **28**, 483.

Von Arx, W. S. (1950). An electromagnetic method for measuring the velocity of ocean currents from a ship under way. *Pap. Phys. Oceanogr. Met.* (M.I.T. and Woods Hole), **11**, no. 3.

Von Arx, W. S. (1961). *An introduction to physical oceanography.* (Addison Wesley.)

Wertheim, G. K. (1954). Studies of the electric potential between Key West and Havana. *Trans. Amer. Geophys. Un.* **35**, 872.

Wollaston, C. (1881). (Tidally induced e.m.f.s in cables.) *J. Soc. Tel. Engrs,* **10**, 50.

Young, F. B. *et al.* (1920). On electrical disturbances due to tides and waves. *Phil. Mag.* **40**, 6th series, 149.

## 8. Other devices, related to induction flowmeters

(Not electromagnetic pumps or magnetogasdynamic generators or accelerators.)

*Electromagnetic brakes*

Baker, R. S. (1960). Design of an eddy-current brake for a sodium-cooled, nuclear power reactor. *Trans. Amer. Elect. Engrs,* Pt. 1, **79**, 330 (formerly U.S.A.E.C. Report NAA-SR-2986 Sept. 1958).

De Bear, W. S. (June 1959). S.R.E. decay-heat problem solved by eddy-current brake. *Nucleonics,* **17**, 108.

Faris, F. E. *et al.* (1958). Operating experience with the S.R.E. *Second U.N. Int. Conf. on Peaceful Uses of At. En. (Geneva).* Paper 452, vol. 9, p. 500.

Schell, F. N. (Aug. 1953). Use of a d.c. electromagnetic pump as a throttling device in a sodium system. *U.S.A.E.C. Report* KAPL-M-FNS-6.

*Liquid metal dynamo for energizing d.c. pump*

Brill, E. F. (1953). Development of special pumps for liquid metals. *Mech. Engng,* **75**, 369.

*Liquid metal pump/dynamo combination*

Jackson, W. D. *et al.* (1961). A magnetohydrodynamic power converter. *Second Symp. on the Eng. Aspects of M.H.D.* (Columbia University Press).

*Induced-field flowmeter*

Lehde, H. & Lang, W. T. (Jan. 1948). Device for measuring rate of fluid flow. U.S. Pat. 2,435,043.

Meyer, R. X. (1961). Some remarks concerning magnetohydrodynamic application to re-entry problems. *Second Symp. on the Eng. Aspects of M.H.D.* (Columbia University Press).

*Force flowmeter*

Shercliff, J. A. (1957). Tests with mercury of a rotary flowmeter for liquid metals. *A.E.R.E. (Harwell) Report* X/M 169.

*Magnetometers*

Kolin, A. (1945). Mercury jet magnetometer. *Rev. Sci. Instrum.* **16**, 209.

Leduc, L. (1887). A new method for measuring magnetic fields. *J. Phys. Théor. Appl.* (2e serie) **6**, 184.

*Ammeter*

Kolin, A. (1945). Mercury manometer ammeter. *Rev. Sci. Instrum.* **16**, 378.

## 9. Incompressible, magnetohydrodynamic channel flow

(Not electromagnetic pumps in general.)

Alpher, R. A. *et al.* (1959–60). Some studies of free surface mercury magnetohydrodynamics. *Bull. Amer. Phys. Soc.* **4**, 282; *Rev. Mod. Phys.* **32**, 758.

Bader, M. & Carlson, W. C. A. (May 1958). Measurement of the effect of an axial magnetic field on the Reynolds number of transition in mercury flowing through a glass tube. *NACA Report* TN 4274.

Birzvalk, Y. A. & Tutin, I. A. (1956). Velocity distribution and magnetohydraulic pressure losses in a rectangular duct. *Trud. Inst. Fiz. Akad. Nauk. Latv. SSSR*, no. 8, 59.

Birzvalk, Y. A. & Veze, A. (1959). Velocity distribution in electromagnetic pump channels with a rectangular cross-section. *Latv. PSR Zināt. Akad. Vēstis*, p. 85.

Boucher, R. A. & Ames, D. B. (1961). End effect losses in d.c. magnetohydrodynamic generators. *J. Appl. Phys.* **32**, 755.

Braginskii, S. I. (1960). Magnetohydrodynamics of weakly conducting liquids. *Soviet Physics JETP*, **10**, 1005 (transl.).

Carlson, A. W. & Sutton, G. W. (Dec. 1960). Effects of end current loops on the velocity profile in a magnetohydrodynamic channel. *General Electric Report* TIS R 60 SD 439.

Chang, C. C. & Lundgren, T. S. (June 1959). The flow of an electrically conducting fluid through a duct with a transverse magnetic field. *Heat Transfer and Fluid Mech. Inst.* p. 41.

Chang, C. C. & Lundgren, T. S. (1961). Duct flow in magnetohydrodynamics. *Z. Angew. Math. Phys.* **12**, 100.

Chekmarev, I. B. (1960). Nonsteady flow of a conducting fluid in a flat tube in the presence of a transverse magnetic field. *Soviet Physics— Tech. Phys.* **5**, 313 (transl.).

Crausse, E. & Poirier, Y. (1957). On laminar flow of an electrically conducting liquid subjected to a transverse magnetic field. *C.R. Acad. Sci., Paris*, **244**, 2694.

Crausse, E. & Poirier, Y. (1957). On turbulent flow of an electrically conducting liquid subjected to a transverse magnetic field. *C.R. Acad. Sci., Paris*, **244**, 2772.

Demetriades, A. (1960). A possible fully developed hydromagnetic pipe flow. *J. Aero-Space Sci.* **27**, 388.

Donaldson, C. du P. (June 1959). The magnetohydrodynamics of a layer of fluid having a free surface. *Heat Transfer and Fluid Mech. Inst.* p. 55.

Drazin, P. G. (1960). Stability of parallel flow in a parallel magnetic field at small magnetic Reynolds numbers. *J. Fluid Mech.* **8**, 130.

Fabri, J. & Siestrunck, R. (1960). Contribution to the aerodynamic theory of the electromagnetic flowmeter. *Bull. Assoc. Tech. Marit. Aero.* no. 60, 333.

139

Fishman, F. (June 1959). End effects in magnetohydrodynamic channel flow. *AVCO-Everett Research Note* 135.

Globe, S. (1959). Laminar steady state magnetohydrodynamic flow in an annular channel. *Phys. Fluids*, **2**, 404.

Globe, S. (June 1959). The suppression of turbulence in pipe flow of mercury by an axial magnetic field. *Heat Transfer and Fluid Mech. Inst.* p. 68.

Globe, S. (1961). The effect of a longitudinal magnetic field on pipe flow of mercury. *Trans. Amer. Soc. Mech. Engrs*, **83**, series C, 445.

Harris, L. P. (1960). *Hydromagnetic Channel Flow*. (Mass. Inst. Tech. Press and Wiley).

Hartmann, J. (1937). Hg-dynamics I. *Math.-fys. Medd.* **15**, no. 6.

Hartmann, J. & Lazarus, F. (1937). Hg-dynamics II. *Math.-fys. Medd.* **15**, no. 7.

Lock, R. C. (1955). The stability of the flow of an electrically conducting fluid between planes under a transverse magnetic field. *Proc. Roy. Soc. A*, **233**, 105.

Lundgren, T. S. *et al.* (1961). Transient magnetohydrodynamic duct flow. *Phys. Fluids* **4**, 1006.

Michael, D. H. (1953). The stability of plane parallel flows of electrically conducting fluids. *Proc. Camb. Phil Soc.* **49**, 166.

Murgatroyd, W. (1953). Experiments on magnetohydrodynamic channel flow. *Phil. Mag.* **44**, 7th series, 1348.

Napolitano, L. G. (1960). On turbulent magneto-fluid dynamic boundary layers. *Rev. Mod. Phys.* **32**, 785.

Pai, S. I. (1954). Laminar flow of an electrically conducting incompressible fluid in a circular pipe. *J. Appl. Phys.* **25**, 1205.

Pavlov, K. B. & Taresov, Y. A. (1960). Stability of flow of a viscous, conducting fluid between parallel planes in a perpendicular magnetic field. *Appl. Math. Mech., Leningr.*, **24**, 1079 (transl.).

Poirier, Y. (1960). Contribution to the experimental study of the magnetohydrodynamics of liquids. *Publ. Sci. Univ. Algér* (Serie B), **6**, 5.

Regirer, S. A. (1960). Flow of electrically conducting fluid in tubes of arbitrary cross-section in a magnetic field. *Appl. Math. Mech., Leningr.*, **24**, 790 (transl.).

Rossow, V. J. (1960). Flow in d.c. electromagnetic pumps. *Rev. Mod. Phys.* **32**, 987. Also *NASA Report* TND 347.

Shercliff, J. A. (1953). Steady motion of conducting fluids in pipes under transverse magnetic fields. *Proc. Camb. Phil. Soc.* **49**, 136.

Shercliff, J. A. (1956). Entry of conducting and non-conducting fluids in pipes. *Proc. Camb. Phil. Soc.* **52**, 573.

Shercliff, J. A. (1956). The flow of conducting fluids in circular pipes under transverse magnetic fields. *J. Fluid Mech.* **1**, 644.

Shercliff, J. A. (1956). Edge effects in electromagnetic flowmeters. *J. Nucl. Energy*, **3**, 305.

Stuart, J. T. (1954). On the stability of viscous flow between parallel planes in the presence of a coplanar magnetic field. *Proc. Roy. Soc. A*, **221**, 189.

Sutton, G. W. (July 1959). Electrical and pressure losses in a magneto-hydrodynamic channel due to end current loops. *General Electric Report* TIS R 59 SD 431.

Sutton, G. W. & Carlson, A. W. (1961). End effects in inviscid flow in a magnetohydrodynamic channel. *J. Fluid Mech.* **11**, 121.

Tarasov, Y. A. (1960). Stability of plane Poiseuille flow of a plasma with finite conductivity in a magnetic field. *Soviet Physics JETP*, **10**, 1209 (transl.)

Uflyand, Y. S. (1960). Flow of a conducting fluid in a rectangular channel in a transverse magnetic field. *Soviet Physics—Tech. Phys.* **5**, 1191 (transl.).

Uflyand, Y. S. (1960). Hartmann problem for a circular tube. *Soviet Physics—Tech. Phys.* **5**, 1194.

Uhlenbusch, J. & Fischer, E. (1961). Hydromagnetic flow in a cylindrical tube. *Z. Phys.* **164**, 190.

Ul'manis, L. Y. (1956). On the problem of edge effects in linear induction pumps. *Trud. Inst. Fiz. Akad. Nauk. Latv. SSR*, no. 8, 81.

Velikhov, E. P. (1959). The stability of a plane Poiseuille flow of an ideally conducting fluid in a longitudinal magnetic field. *Soviet Physics JETP*, **9**, 848 (transl.).

Williams, E. J. (1930). The induction of e.m.f.s in a moving liquid by a magnetic field and its application to an investigation of the flow of liquids. *Proc. Phys. Soc., Lond.*, **42**, 466.

Woodrow, J. (1949). The d.c. electromagnetic pump for liquid metals. *A.E.R.E. (Harwell) Report* E/R 452.

Wooler, P. T. (1961). Instability of flow between parallel planes with a coplanar magnetic field. *Phys. Fluids*, **4**, 24.

10. *Induction flowmeters. Review articles and miscellaneous*

Balls, B. W. & Brown, K. J. (1959). Magnetic flowmeter. *Trans. Soc. Instrum. Tech.* **11**, 123. See also: Foxboro magnetic flowmeter—*Blast Furnaces and Steel Plant*, **42** (1954), 1332. *Machine Design*, **28** (1956), 98. Calibration tests of 8 in. magnetic flowmeter—(Apr. 1955). *Alden Hydraulic Lab. Report, Worcester (Mass.) Polytechnic*. Brit. Pat. 789,102 (Sept. 1955).

Boeke, J. (1953). A new method of flow measurement. *Chem. Weekbl.* **49**, 133.

Boeke, J. (1953). Flow measurement on a basis of electromagnetic induction. *Electronica*, **6**, 81. See also: Tobiflux electromagnetic flowmeters. *Engineering, Lond.*, **176** (1953), 542. Brit. Pat. 726,271 (May, 1953).

Fisher, J. H. (June 1952). Investigation of an electromagnetic method for measuring fluid flow rates. *U.S.A.E.C. Report* AD-47751.

Fisher, J. H. (1955). Electromagnetic flowmeter design and performance considerations. *Trend Engng. Univ. Wash.* **7**, 18.

Fukawa & Sakata (1955). Report on electromagnetic flowmeter manufactured by trial. *J. Soc. Instrum. Tech. Japan*, **5**, 393.

Gast, T. & Vieweg, R. (1950). Remote measurement of flow by electrical instruments. *Feinwerk Technik*, 54, 261.

Gray, W. C. & Astley, E. R. (June 1954). Liquid metal magnetic flowmeters. *Instrum. Soc. Amer. J.* 1, 15.

Head, V. P. (1959). Electromagnetic flowmeter primary elements. *Trans. Amer. Soc. Mech. Engrs*, 81, series D, 660.

Horiuchi, T. B. & Shibata, T. (1959). Stability of electromagnetic flowmeters. *Proc. Fujihara Mem. Fac. Engng, Keio Univ.* 12, 46.

Hutcheon, I. C. (Apr. 1960). Some problems of magnetic flow measurement. *Instrum. Engng*, 3, 1.

Kamp, J. J. & Smals, J. L. (1954). Electromagnetic flowmeter for industrial application. *Polyt. Tijdschr.* A, 9, 530.

Katys, G. P. (1956). Apparatus for the measurement of a rapidly changing discharge of liquid. *Izmerit. Tekhnika*, 4, 87.

Katys, G. P. (1957). Flowmeters of small inertia for the measurement of pulsating flows. *Priborostroenie*, 8, 29.

Kolin, A. (Mar. 1939). Apparatus for measuring fluid flow. U.S. Pat. 2,149,847.

Linford, A. (1961). *Flow Measurement and Meters.* (2nd ed.) (Spon), ch. 7.

Munch, R. H. (Jan. 1952). Magnetic induction flowmeter (Magnaflow). *Industr. Engng Chem. (Industr.)*, 44, 83 A; *Chem. Engng*, 59, 192.

Northfield, H. J., Howe, G. W. O. & Taylor, J. E. (1936). Fluid flow past magnet poles. *Electrician*, 116, 14, 40, 68, 96, 151, 208.

Raynsford, C. K. (Dec. 1948). Suitability report on electromagnetic flowmeters. *U.S.A.E.C. Report* NP-6520.

Regnart, H. C. (1930). Generation of electric currents by water moving in a magnetic field. *Proc. Univ. Durham Phil. Soc.* 8, 291.

Remeniéras, G. & Hermant, C. (1954). Electromagnetic measurement of speed in liquids. *Houille Blanche*, 9, 732.

Robin, M. J. (1953). Electromagnetic flowmeter. *J. Rech.* 5, 187.

Savastano, G. & Carravetta, R. (1954). Electromagnetic apparatus for measuring the speed of flowing liquids. *Energia Elett.* 31, 81.

Shcherbakov, G. (1958). An electromagnetic flow detector. *Radio (Moscow)*, 7, 27.

Shercliff, J. A. (1955). Some engineering applications of magnetohydrodynamics. *Proc. Roy. Soc.* A, 233, 396.

Soffler, A. (Aug. 1953). Manufacturing specification. Magnetic induction flowmeter. *U.S.A.E.C. Report* KLX-1392.

Spink, L. K. (1958). *The Principles and Practice of Flowmeter Engineering.* (8th ed.) Foxboro Co.

Thornton, W. & Bejack, B. (1959). Performance and applications of a commercial blood flowmeter (Avionics). *Trans. Inst. Radio Engrs*, ME-6, 237.

# INDEX

# CONTENTS

CHAPTER VI

## Radiation

# AUTHOR'S PREFACE

I wrote this monograph assuming that the reader would be a physicist who was interested to learn something of the progress in a subject outside his own field of study. I should point out therefore that there is a very great difference between the approach to a laboratory problem and the approach to a geophysical (or, for that matter, astronomical) problem. In geophysics, and particularly in atmospheric physics, experiments are either impracticable or of limited value. Controlled conditions are essential for a successful experiment, while with the atmosphere it is only possible to observe and to hope that the conditions relevant to a unique interpretation are either known or can be guessed. It is not surprising that this may lead to incompatible results from two apparently reliable observations, each intelligently interpreted, and there are examples of this kind in the following pages. While the laboratory worker would wisely reject such results as untrustworthy, the atmospheric physicist must often accept them as the only results he is likely to obtain.

Every research has its own difficulties, and atmospheric physics is not unique in the effort, expense and even personal discomfort that may be involved in gathering representative observations. However, once again, it differs from most branches of physics in working with a system which is very far from equilibrium. As a result it is very difficult to specify just how many observations are desirable, and, moreover, when it comes to theoretical interpretation, equilibrium hypotheses, which are normally so valuable, turn out to be useful for a rough superficial examination only.

In presenting this subject to those who have not studied geophysics, I have sought to emphasize characteristic features where they appeared to be instructive, but to avoid issues which seemed to me to be particularly confusing. If this monograph had been written for the geophysicist, I would have had to consider in greater detail the differences of opinion which exist upon nomenclature, the relative importance of different topics, the relative reliability of different observations, etc. The meteorologist, for example, might well consider that my treatment of dynamical matters is most inadequate. I hope, however, that those who specialize in this subject will derive some of the interest from reading this monograph that I have derived from writing it.

I am indebted to Drs G. K. Batchelor, M. V. Wilkes and T. W. Wormell for commenting upon some sections. Mr C. D. Walshaw read the manuscript with care and found many minor errors and obscurities. Mr H. E. Goody has helped me greatly with his thorough reading of the proofs.

R. M. GOODY

*Cambridge* 1953